Skeptoid 5

Massacres, Monsters, and Miracles

By Brian Dunning

Foreword by Ronald Hayden
Illustrations by Mitsuko Stoddard

Skeptoid Podcast ©2006-2013 by Brian Dunning
http://skeptoid.com

Published by Skeptoid Media
Laguna Niguel, CA

First Edition
ISBN: 978-1492709060
Printed in the United States of America

The true delight is in the finding out rather than in the knowing.

Isaac Asimov

ACKNOWLEDGEMENTS

This book was made possible through the kind support of:

Ronald Hayden
Brian Walker
John Yockachonis

Thank you, friends, for your generous trust in this project and your contribution to science education.

To Lisa – the light in my day, the moon in my night, the chocolate in my sauce, the twinkle in my eye.

Contents

Foreword: Zombies and Dolphin Combat 1

Introduction: Why I Love Science............5

1. Finding the Lost Colony of Roanoke 9

2. The Miracle of Calanda............17

3. The Exorcism of Anneliese............23

4. The Voynich Manuscript............30

5. The Port Arthur Massacre............37

6. Finding the POW/MIAs............43

7. Superhuman Strength during a Crisis............50

8. The Secret of Plum Island 56

9. Spontaneous Human Combustion............63

10. The Secret History of Chinese Medicine 69

11. Military Dolphins: James Bonds of the Sea 76

12. Near Death Experiences 83

13. The Haitian Zombies 90

14. Anastasia............97

15. Conspiracy Theorists Aren't Crazy............104

16. "Curing" Gays............110

17. Zeno's Paradoxes............117

18. The Abominable Snowman 124

19. The Hessdalen Lights............131

20. The Zionist Conspiracy 138

21. Are We Alone? 145

22 All About Fracking 152

23. The Monster of Glamis......159

24. Brainwashing and Deprogramming......167

25. Noah's Ark: Sea Trials......174

26. Finding Shakespeare......181

27. The Science of Voting......188

28. The Jersey Devil......194

29. Top 10 Worst Anti-Science Websites......201

30. The Fate of Fletcher Christian......211

31. Korean Fan Death......218

32. Pit Bull Attack!......225

33. The Mystery of the Mary Celeste......232

34. Approaching a Subject Skeptically......239

35. The Toxic Lady......245

36. The Grey Man of Ben MacDhui......252

37. Wunderwaffen: Nazi Wonder Weapons......259

38. Finding Amelia Earhart......266

39. The Versailles Time Slip......274

40. A Magical Journey through Reasoning Errors......282

41. Star Jelly......289

42. The Beale Ciphers......295

43. De Loys' Ape......302

44. Are Vinyl Recordings Better than Digital?......309

45. Catching Jack the Ripper......316

46. I Can't Believe They Did That: Human Guinea Pigs....323

47. The Siberian Hell Sounds......331

48. Picnic at Hanging Rock......338

49. The Science and Politics of Global Warming.................345

Epilogue...358

FOREWORD: ZOMBIES AND DOLPHIN COMBAT

BY RONALD HAYDEN

Let me tell you a secret. I don't always agree with Brian Dunning.

In fact, if we ever get a chance to have a drink together, he's going to get an earful on a couple of those topics where his conclusions are out to lunch. In my imagination, by the end of the evening he has succumbed to my superior logic and experience and has agreed to rewrite the *Skeptoid* episodes in question; in reality, I have a pretty good idea who would actually be set straight, especially given his annoying habit of falling back on research and facts and the like.

But here's the thing: I cherish these occasional disagreements. They are a good thing. If they didn't occur, I might well have stopped paying attention to *Skeptoid* by now.

When I was a young lad, over and over I would latch onto a potential "guru" who seemed to agree with me in all things. I would bask in their obvious correctness and happily accept whatever they said with the warmth that comes from knowing you are in complete synchronicity with someone...until, each and every time, I discovered one critical thing on which we disagreed, at which point I would immediately drop that guru and go through a period of massive disappointment and emotional turmoil. How could someone who agreed with me so much be SO WRONG? It just didn't seem possible.

I believed that all right-thinking people had all the same right thoughts.

As an older lad, I've gone through the looking glass. The more I find myself agreeing with someone, the more antsy I get. It means the answers — and the questions — are too easy. It means I'm letting myself get intellectually lazy. The next step is to devolve into one of those self-congratulatory conversations where we reflexively confirm each other's unexamined beliefs while smugly trashing anyone with a different point of view.

Ugh, the next morning I hate myself for those conversations.

Nowadays, it's at that moment when I find our area of disagreement that I relax and engage. That's when I know there's value to be had in the conversation, that I can trust myself because knowing it's possible for us to disagree, my brain refuses to let down its defenses. I'm going to examine each claim on its own merits and make my own decision.

So my thanks to Brian for being someone I only agree with 99% of the time. The 1% keeps it interesting.

And my thanks for his being that all-too-rare skeptic who remembers that skepticism isn't about taking the fun out of life. It's not just glumly and smugly letting the world know for the 2,010th time that ghosts don't exist and that crop circles aren't from UFOs. Skepticism is the excitement of exploring the universe and discovering what's actually out there, which is always more exciting and mysterious than the stories we make up about what's out there.

Understanding this, Brian goes wherever his curiosity takes him. So in this book he has essays on topics ranging from zombies to fracking to dolphins in combat. Dolphin combat, people!

No matter who you are, no matter how long you've followed science or the skeptical community, many if not most of these essays will surprise and delight you. And just maybe you'll find one where Brian has slipped up on a minor fact or come to a conclusion you don't agree with.

That's when the fun starts.

Ronald Hayden is a former amateur magician who grew up on classic science fiction. Both these factors led naturally to an interest in skepticism. He has worked for 20 years at a Silicon Valley technology company, where he now manages a publication department.

INTRODUCTION: WHY I LOVE SCIENCE

I was 10 or 12 years old when we were riding in a VW bus through the mountains. There were half a dozen of us, my brother, my dad, and some of his friends. Shotguns were loaded in the back, and we were on our way to a campground, I wasn't quite sure where, and it was hunting season for some animal I didn't quite know. I spent most of the ride with my nose pressed to the window, looking through the trees. The woods were unusually dark and dense; even the occasional meadows seemed closed-in and a little scary.

When I saw the first of them, it gave me a bit of a rush but I don't remember it being particularly scary. It seemed a natural enough looking animal, something that belonged there. But it struck me as very unusual to see.

It was only after I saw a second and a third, very clearly out in one of the meadows, looking at us unconcernedly, that I was hit with the first pang of fear. *This is real,* I thought. *This is no dream. This is happening. There are Sasquatches all around us, and we're in grave danger.*

When I cried out and told everyone in the car what was out there, they didn't seem to care. I jumped wildly from window to window. A Sasquatch was to be seen in virtually every direction, stepping casually over a log, walking along the side of the road, shaking a tree. I pointed them out specifically. Nobody could have cared less. My companions hardly responded at all. Sasquatches aren't real; it was scarcely even worth acknowledging the small boy's cries. They weren't even concerned when one crossed the road right in front of us; they just drove blindly on, continuing their conversation, aloof to the monsters outside.

This has happened to me many times. It's a recurring dream that I've had most of my life. Dreams offer no relief from the virtual reality they present. The emotions they trigger are every bit as real as those in the waking world. There is no refuge in the realization that the images we dream are only phantoms, for when dreams happen, they are all too real. Despite having no concern about Bigfoots in the comfortable light of day, I can honestly say that I know the horror of coming face to face with one.

This aptly illustrates the value of myths, urban legends, and popular pseudosciences. It's not the subject of the legend that holds the wonder; it's the science behind how and why the legend exists, and its sociological manifestations. There's nothing to be found of any interest if you treat Bigfoot as a real animal and go looking for it; it isn't there. But there is a lot to be learned from dreams, for example; how and why they form, what need does the brain have that the dream fulfills.

When a diet fad spreads through society like wildfire, taking over the store shelves and the daytime talk shows, there is rarely anything science-based to the diet itself. But the life cycle of the fad teaches us about psychology and marketing. Believing the claims made by the diet's proponents will rarely do you any good, but understanding the reasons for the fad will indeed make you better able to navigate through life.

You're making a documentary film about a ghost said to inhabit a particular hotel room. You rent the room, set up your camera and infrared gear and time lapse, and… nothing happens. There is nothing to be learned from belief in a legend; but there is in the study of the legend. If you'd gone down and interviewed the bartender or the head proprietor about how they first heard of the story, you'll get a real history lesson and learn about real people. If you're lucky, you might even find out about a perceptual phenomenon that tricks us into seeing, hearing, or feeling something that isn't there.

You have a young daughter and your pediatrician tells you that it's time for scheduled vaccines. You've heard on the television and from friends that vaccination carries more risk than benefit. What do you do? Should you believe the popular anecdotes, or should you understand the science behind the recommendation? Often in life, the stakes are higher than just whether or not you choose to believe in a ghost or spend some time throwing money at a fad diet. Sometimes when we decide whether to believe what's popular or to take a science-based view, it's a matter of life and death.

Applying science to our daily encounters has proven richly rewarding for me, time and time again. One subject I'll discuss in later chapters is the Abominable Snowman, which for most people, is a question of whether or not there's a snow monster roaming around the Himalayas. But that's the wrong question to ask. Instead, how do we know whether there is (or not) an Abominable Snowman? The study gave me unexpected insights into both psychology and zoology.

I'll discuss topics as varied as the science of voting. Countless statisticians and game theorists have, for thousands of years, sought ways to sway free elections, and surprising theorems exist. The one that most opened my eyes is that, when there are more than two candidates and you're only allowed to vote for one (as we do it here in the United States), there is no fair way to hold the vote: blocs can successfully control the outcome to hand the victory to a less popular candidate; and even when they don't, the Condorcet candidate (the one who would beat all others in a head-to-head election) is often not the winner. It's a fundamentally broken system, but it turns out there are at least a couple of possible easy fixes to assure that the Condorcet candidate will nearly always win.

Basic science literacy and a grasp of critical thinking methodology certainly do provide real-world applications. But that's only one reason I love science. The main reason is simply the rush of excitement from learning something new and amazing. Every topic I've ever covered on my show has given me the gift of a marvelous tidbit. There's a unique satisfaction to be gained from learning a new fact in which you can have confidence in its ability to withstand scrutiny – a satisfaction certainly *not* to be found in the popular supernatural explanations that bombard us all day long.

I hope you enjoy these chapters. Be prepared for two things: First, that you will be as excited as I was to find out some crazy details about our world; and second, that there will probably be at least one sacred cow of yours in here for which I'll find an alternate explanation. That's the nature of the journey: you never quite know what's around the next corner.

– Brian Dunning

1. FINDING THE LOST COLONY OF ROANOKE

One of the early attempts of the English to colonize America ended with every person simply disappearing from the Roanoke colony.

When a relief party sent from England arrived at the Roanoke Colony on the East coast of the United States in 1590, they found the settlement neatly dismantled, and not a soul to be found. Some 115 men, women, and children had simply disappeared. The only clue was the name of a nearby island, Croatoan, carved into the trunk of a tree. And thus was launched one of history's great mysteries: The Lost Colony of Roanoke.

Around 1580, the English mogul, explorer, and all-around famous guy Sir Walter Raleigh made a deal with Queen Elizabeth I to establish an English colony in America. He was given ten years to do it, and the deal was that they'd share in the riches they hoped would be found, and also create a base for English ships fighting

Spain. Raleigh did not personally go, but his first expedition sailed in 1584 to find a good location.

And find one they did: Roanoke Island, off of North Carolina. It's located inside the outer banks, a long string of narrow sand spit islands that shelter half the coast of North Carolina. Roanoke is fertile, defensible, well wooded, and offers substantial protected anchorage for ships. But Raleigh's expedition made some poor choices. Among the first things they did was to pick a fight with the local natives, charging them for some petty theft, burning their village, and beheading their chief. This was perhaps not the best way to establish friendly relations.

Sir Francis Drake happened by during some of his pirating exploits, and finding the men in poor condition, he gave them a lift back to England. Strike 1 for the New World colony. What nobody knew was that Raleigh's second ship was already on its way. The two ships passed each other in the Atlantic, and the new group found an abandoned settlement. They returned to England, but left a small garrison of fifteen men on Roanoke to protect Raleigh's claim. The garrison was unaware of the bad diplomacy they'd inherited, and it should come as no surprise that they were never seen or heard from again. Strike 2 for the New World colony.

Raleigh sent a third expedition to Roanoke in 1587, larger and better provisioned than the predecessors, commanded by Roanoke veteran John White, a prolific artist and cartographer who had originally been hired to document the colonization through his artwork. White re-established the Roanoke settlement, but failed to rebuild relations with the natives. They sometimes skirmished, and being vastly outnumbered and completely on their own, feared for their lives. As this was the first colony to include women and children, including White's daughter and her family, he was persuaded to return to England with a skeleton crew to ask Raleigh for help. With relocating the colony as an acknowledged possibility, White left instructions that if the colonists did choose to move in his absence, to carve their destination on a certain tree; and that if they were in trouble, to also carve a Maltese cross. Arrived in England, White found that the war with Spain had complicated matters, and it was three long years before he was finally able to return with an armed party and the needed supplies.

Unfortunately, there was nobody on Roanoke to receive them. The camp had been tidily dismantled; there had been no sudden massacre. White went to the tree and found a single carved word, Croatoan, and no carved cross. Wherever they'd gone, their departure appeared to have been orderly and planned, and they had not been in any immediate danger.

Croatoan was a barrier island on the outer banks, now called Hatteras Island, about 35 nautical miles south of Roanoke. Today its flat dunes are covered with hurricane-hardened vacation homes. Sport fishing boats come and go,

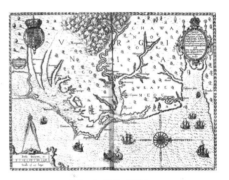

Map by John White

and kite boarders take advantage of its strong winds. But in the colonial days, it was one home of the Croatan natives, who were friendly to the English. It would seem to have been a logical destination, had food run out on Roanoke or if there had been some other cause to leave.

Unfortunately, the weather had plans for John White that did not include allowing him to make the short hop to Croatoan to find his colony. A storm came in just as they arrived at Roanoke, and White's ship lost its main anchor. The combination of storm waves and wind, and a lost anchor, made it impossible to safely navigate the coastal islands and to land anywhere. The captain of the ship hired by White was anxious to get back to the more profitable business of privateering against Spain, and rather than his risk his vulnerable ship in a dangerous and fruitless coastal search, he opted to return to England. White arrived home empty handed. It was Strike 3, and nobody ever again heard of Sir Walter Raleigh's lost colony of Roanoke. It was the end of their story, and also the beginning of their legend.

But colonies ultimately did take root in America, and it seems that those early colonists closest to Roanoke may have learned something of what became of them. The lost colony did eventually have neighbors, but they came some 25 years later and some 125 nautical miles of sailing to the north. It was 1607 when a more permanent colony was finally established in Virginia: Jamestown, named after King James, later famous for Captain John Smith, Pocahontas, and John Rolfe's tobacco. Jamestown struggled badly in its first years, and most of its colonists died from starvation or disease. They had few resources and never mounted an expedition to Croatoan to see what had become of their predecessors.

Jamestown was in Powhatan native territory. Shortly before the Jamestown colonists had arrived, or at about the same time, the Powhatan had attacked and exterminated the Chesapeake (yes, the lovely Pocahontas was from a genocidal tribe). If any English or half-English had survived and been living with the Chesapeake, Powhatan killed them too. Relations between the English and the Powhatan were frosty, but the English were able to gain some small amount of information about what had become of the Roanoke colony.

According to the Powhatan who were willing to talk, the original colonists had integrated into the mainland Carolina tribes to the west of Croatoan. There were a number of stories that supported this. John Smith was told there was a town where men dressed as he did, and another Englishman, William Strachey, wrote that he was told of:

"...howses built with stone walles and one story above another, so taught them by those Englishe who escaped the slaughter at Roanoak, ...where the people breed up tame turkeis, about their howses, and take apes in the mountaines, and where, at Ritanoe, the Weroance, Eyanoco preserved seven of the English alive, - fower men, two boyes, and one yonge mayde (who escaped and fled up the river of Chanoke), to beat his copper of which he hath certaine mynes."

Copper working was indeed known to the Native Americans at that time. Further evidence for these stories was found later by Spanish agents, whose job it was to eliminate evidence of the English occupying and possessing America. They recovered a chart drawn by a Jamestown man, now known as the Zuniga Map, and sent it to Europe in 1608. The chart was inscribed with the place names mentioned by Smith and Strachey, but also mentions "four clothed men from Roonok."

There is only one account of the Jamestown colonists actually encountering a person who appeared to be of English descent. He was seen in Powhatan territory in 1607, and was described as:

"a Savage Boy about the age of ten yeeres, which had a head of haire of a perfect yellow and a reasonable white skinne, which is a miracle amongst all the Savages."

If this is a true account, and the witnesses were not mistaken in their observation and reporting, then this boy would have been born some seven years after John White found the deserted colony. It seems probable that the boy was a descendant of the Roanoke settlers who had either been spared by the Powhatan or was raised by them. Either way, it would be evidence that some of the Roanoke colonists did seek refuge to the north.

Other accounts place the colonists south, either on Croatoan or on the Carolina coast. A century later, surveyor John Lawson wrote that the Hatteras natives on Croatoan:

> *"tell us that several of their Ancestors were white People, and could talk in a Book, as we do; the Truth of which is confirmed by gray Eyes being found frequently amongst these Indians, and no others."*

The late Irish historian David Beers Quinn was probably our most knowledgeable scholar on the Roanoke colony. He devoted his career to the study of the colonization of America. Quinn's theory, developed after decades of studying many such accounts, is that the colonists abandoned Croatoan and separated into two groups. One group peacefully assimilated into the Carolina tribes to the west, and the other group went north to live with the friendly Chesapeake, reasoning that Virginia was the most likely place to which any future English colonists might come. We know that any who went north ultimately died, either naturally or at the hands of the Powhatan, but the fate of those who went to the Carolina mainland is less clear.

Quinn's theory is the best available based on historical study, and needs only archaeological and genealogical evidence to back it up. Fortunately, both are forthcoming — sort of. There is archeological evidence that the colonists may have lived for a time on Croatoan, though it's not conclusive. Alongside artifacts of the Croatan natives, a gold ring was found bearing the family crest of the Kendall family, and there was a Master Kimball in the original Roanoke colony. However two English copper farthings were also found on Croatoan, but they were not produced until the 1670's, nearly a century too late to have belonged to anyone who was in the colonies at the time of Roanoke.

The DNA evidence supporting Quinn's theory may prove to be more revealing. Several different groups throughout the United States are currently analyzing DNA test results looking for clues to whether the colonists' genes survive today. Native populations at the time were quite small, and so it's more than likely that intermarriage would leave a substantial genetic footprint. In an article published in 2005 in the *Journal of Genetic Genealogy*, researcher Roberta Estes gave a thorough rundown of what's known so far, and unfortunately, the strongest point made is that there is insufficient data.

When I think of the lost colony, I think of John White; of his many beautiful illustrations of native life in the Americas, and of the anguish he felt at sea having left his daughter at Roanoke for more than three years. I think of how helpless he must have felt, being unable to find ships to return. I think of how it must have been to find the word Croatoan carved into the tree, of his doubtless desperation to continue, and the frustration of having to return to England when he was so close to finding them. Where were Eleanor White, her husband, and White's infant granddaughter? Were they living safely with the Croatans, or had they succumbed to disease or starvation? He never knew, and we'll never know either. We may find better evidence of what happened to the colony as a whole, but the small personal tragedies will always remain unaccountable.

References & Further Reading

Baxter, J. "Raleigh's Lost Colony." *The New England Magazine.* 31 Dec. 1894, Volume 11: 565-587.

Estes, R. "Where Have All the Indians Gone? Native American Eastern Seaboard Dispersal, Genealogy and DNA in Relation to Sir Walter Raleigh's Lost Colony of Roanoke." *Journal of Genetic Genealogy.* 1 Jan. 2009, Volume 5, Number 2: 96-130.

Miller, L. Roanoke: *Solving the Mystery of the Lost Colony.* New York: Arcade, 2001.

Quinn, D. *The Roanoke voyages, 1584-1590; documents to illustrate the English voyages to North America under the patent granted to Walter Raleigh in 1584.* London: Hakluyt Society, 1955.

Rountree, H. *Pocahontas's People: The Powhatan Indians of Virginia through Four Centuries.* Norman: University of Oklahoma Press, 1990.

Stahle, D., Cleaveland, M., Blanton, D., Therrell, M., Gay, D. "The Lost Colony and Jamestown Droughts." *Science.* 1 Apr. 1998, Volume 280, Number 5363: 564-567.

2. THE MIRACLE OF CALANDA

They say God doesn't heal amputees... but apparently he did, once. Or did he?

A favorite question asked by skeptics, when confronted with stories of miraculous religious healings, is to ask "Why doesn't God heal amputees?" The answer? He did, once. It happened in Spain in 1640, when a young man's injured leg was amputated. Two and a half years later, his leg was miraculously restored. It's become known as the Miracle of Calanda, and it's perhaps one of the best documented of miracles. The faithful have hard evidence to back it up, and the skeptics have no answer. Was the event truly miraculous and unexplainable? Maybe it was; but we're going to take a hard look at what's actually known, and see if we can uncover the most likely explanation.

Miguel Juan Pellicer, a strapping young fellow about 20 years old, was working at his uncle's farm in the village of Castellón in 1637. A mule-drawn cart ran over his leg, fracturing the tibia. Quickly, his uncle drove him to the hospital at Valencia. The story, as recorded, says that Pellicer stayed in the Valencia hospital for five days, until it was decided that he needed better help than they could provide. Pellicer was sent, on foot, with a broken leg, to the larger hospital in Zaragoza, a journey which took him 50 days.

Once he arrived in Zaragoza, feverish and ill, doctors found his leg to be gangrenous and in a grievous state. Pellicer's right leg was amputated "four fingers below the knee" and it was buried in a special plot at the hospital. He stayed in the hospital for several months, and was provided with a wooden leg and a crutch. He then applied to the church authorities at the Basilica of Our Lady of the Pillar in Zaragoza for authorization to make a living as a beggar, which was granted. Pellicer lived in Zaragoza for two years, attending mass daily at the Basilica, and accepting alms from the citizenry. The pious young amputee was a familiar face in town.

At last he decided to return home. He rode a donkey all the way to his parents' home in Calanda, where he'd grown up. His family was overjoyed to see him, but since he couldn't work, he spent a couple of weeks riding his donkey to neighboring villages begging. And then one night, it happened.

Basilica at Zaragoza

A traveling soldier was spending the night in Pellicer's own room, so Pellicer took a bedroll on the floor in his parents' bedroom. In the morning, his parents saw not one, but two feet protruding from the end of the short blanket! They excitedly woke their son, who was as surprised as anyone, and the news quickly spread throughout the village that the young amputee had been miraculously healed.

An examination of the leg revealed it was the same leg he'd always had. It bore a scar from where a cyst had been excised when he was a child, two scars made by thorns, and another from a dog bite on his calf. Most notable was a scar where the cart wheel had crushed his tibia. The leg was said to appear thin and atrophied, but within a few days he was using it normally.

As the story spread, it drew in the curious and the official. A few days after the miraculous restoration, a delegation consisting of a priest, a vicar, and the local royal notary came to Calanda to see for themselves and to prepare an official record of the event. They took statements from witnesses and carefully documented Pellicer's story.

Two months later, a trial was opened in Zaragoza where more than 100 people testified that they had known Pellicer with only one leg, whereas now he had two. Ten months later, the archbishop rendered a verdict that the restoration of the leg was canonized as a true miracle. Since that date, skeptics have no longer been able to charge that God does not heal amputees.

The most authoritative work on the Miracle of Calanda is the 1998 book *Il Miracolo* by Catholic scholar Vittorio Messori which identifies and records the pieces of written evidence collected by the delegation, and which survives today. They are the following:

- Documentation of Miguel Juan Pellicer's baptism, confirming that he was a real person.

- Registration of Pellicer's admittance to the hospital at Valencia.

- The delegation's original notarized report of the statements collected in Calanda, including statements by people who saw him come to town with one leg and wake up with two.

- A certified and notarized copy of the original minutes of the trial at Zaragoza, including many statements of people who knew Pellicer as a one-legged beggar.

There are also many other documents that do not necessarily support the miracle claim, but that support other parts of the story; for example, proof that other people named in the story exist, proof that after the miracle Pellicer was invited to the royal court in Madrid, and books and other publications retelling the event.

If we accept that these documents are indeed legitimate, and I think we can, is there any wiggle room left? Do the documents consist of proof that a miraculous restoration of an amputated limb occurred?

Medically, Pellicer's story is improbable, but not impossible. 55 days after the injury, he said, his leg was amputated due to advanced gangrene. In a crushing injury like the one he suffered, gangrene may take from 48 to 72 hours to set in, and once it does, you're gone from sepsis in as little as a few hours. Nobody lives 55 days with a gangrenous injury. If his skin was not broken, or if any breaks healed cleanly, it is still possible that the wound could have developed internal gas gangrene weeks, months, or even years later. But the appearance of gas gangrene is inconsistent with the condition allegedly reported by the doctors, which was "phlegmonous and gangrenous", meaning open and wet, and "black". Without an actual examination, we can't say for certain that Pellicer's story is impossible; but the version of the story that's been reported raises a huge medical red flag.

This red flag is sufficient to prompt a closer examination of the documented evidence. And there is one thing that jumps out. It's a giant, gaping hole. In case you haven't fallen into it yet, or seen any large buildings or 747s get swallowed up in this hole, I'll point it out: *There is no documentation or witness accounts confirming his leg was ever gone.*

But what about all those witnesses who knew him with one leg? Allow me to offer an alternative version of what might have happened, that requires no miraculous intervention, and is still consistent with all the documentary evidence we have. Pellicer's leg was broken in the accident as witnessed and reported, but like most broken legs, did not develop gangrene. His uncle took him to the hospital at Valencia (a documented event), where he spent five days — during which his uncle presumably went back to his farm — and his broken leg was set.

The next 50 days he spent convalescing as his leg mended. Unable to work during this time, he was forced to earn a living as a beggar, and found that the broken leg did wonders for the collection of alms. Once his leg was sound, he reasoned that if a broken leg was good, a missing leg would be even better. He bound his right foreleg up behind his thigh, got ahold of a wooden leg, and traveled to Zaragoza, home of the great Basilica — someplace where he wasn't known. For two years, the young Pellicer enjoyed the relative financial success of panhandling among the Basilica's devotees as an amputee with a sad story.

Eventually he made it back home to Calanda, where his plans were accidentally foiled when the existence of his complete, sound leg was revealed when his parents saw his feet sticking out of his blanket. At that point, the miracle story was a perfect cover. Many, many people had known him as the man with one leg, and now everyone could quite plainly see that he had two. There was no way he could lose.

I'm not accusing Miguel Juan Pellicer of being a fraud, but I am pointing out that there is a far more probable alternate explanation. Faking blindness, infirmity, poverty, and all manner of ailments is hardly unheard of among beggars. It is now, and has been for millennia, a pillar of the profession.

Note that no evidence exists that his leg was ever amputated — or that he was even treated at all — at the hospital in Zaragoza other than his own word. He named three doctors there, but for some reason there is no record of their having been interviewed by either the delegation or the trial. The trial did find that no leg was buried where he said it was at the hospital, but this is exactly what we'd expect to find if it had never been amputated. Although this lack of a buried leg is often put forth as evidence that the story is true, it is actually a lack of evidence of anything.

We have evidence that he was admitted to the hospital in Valencia with his uncle. We have notarized first-hand statements that a scar was visible on his leg where he had been injured by the mule cart. We have numerous statements that he was well known in Zaragoza as a one-legged beggar. All the evidence supports Pellicer being a beggar with a popular and time-honored gimmick who was caught, not with his hand in the cookie jar, but with his feet out of the blanket. It is only through the introduction of a new assumption, that of the existence of unprecedented supernatural intervention, can the alternate explanation of a miraculous restoration be found consistent with this same evidence. This is where Occam's Razor comes into play: The most likely explanation is the one that requires the fewest new assumptions.

We can't say that the Miracle of Calanda is not genuine, and we can't prove that Miguel Juan Pellicer's leg was not miraculously restored. But we can say that the evidence we have falls short, and is perfectly consistent with no miracle having taken place.

REFERENCES & FURTHER READING

Aranda, J. *Luis Buñuel: A Critical Biography.* London: Secker and Warburg, 1975. 7-17.

Domingo Pélrez, T. *El Milagro de Calanda y sus Fuentes Históricas.* Zaragoza: Caja Inmaculada, 2006.

Marie, A. "Spiritual Newsletter." *Saint Joseph de Clairval Abbey in Flavigny.* Abbey of Saint-Joseph de Clairval, 8 Dec. 2006. Web. 23 Feb. 2011.
<http://www.clairval.com/lettres/en/2006/12/08/2061206.htm>

Naval, L. *El Milagro de Calanda a Nivel Histórico.* Zaragoza: Iglesia Católica, 1972.

Sanz y Martínez, M. *Calanda.* Reus: Artis Gráf, 1970. 1-41.

Vittori, M. *Il Miracolo.* Milano: Rizzoli, 1999.

3. The Exorcism of Anneliese

Can exorcisms such as the one that killed Anneliese Michel truly help critically ill people?

It doesn't just happen in the movies. Throughout the centuries and in all countries, the faithful have practiced exorcism. It's a religious ritual intended to drive away demons who are possessing a victim's body. Its basic premise — that any such thing as demons or demonic possession even exist — places it outside the bounds of what can be tested or evaluated, let alone proven. But this has not stopped it from being employed in life-or-death situations: Medical emergencies entrusted to prehistoric superstition. Can exorcism truly treat severely disturbed individuals?

This was the question with Anneliese Michel, a young Bavarian woman born in 1952. Anneliese was raised in a profoundly devout Catholic home; three of her aunts were nuns and her father had studied to become a priest. But the Michel home had a profane secret: An illegitimate daughter, Martha, born four years before Anneliese. Martha died of kidney trouble when Anneliese was still a child, and compounded with the shame of the illegitimate birth, it rocked the pious family to its core. Anneliese performed constant penance for her sinful mother. The family turned to fringe extremist Catholic groups, forging their own form of deep religious piety. Anneliese's upbringing was one quasi-Catholic rite after another, a constant atonement for the sins of others.

At the age of sixteen she began having epileptic seizures. For the next few years, she was in and out of psychiatric hospitals, on and off of a half dozen antipsychotic and antiepileptic drugs, and her behavior spiraled worse and worse. Anneliese became obsessed with atonement and ritual, but it went much further than that. She reported visions of demonic faces and panicked and snarled at sacred images, and the seizures continued and became more bizarre. After exhausting all the medical options, the Michel family turned to the church for help. Over the final year of her life, Anneliese received no medical care (at her own demand) and was put through sixty-seven exorcism sessions, as codified in Roman Ritual. Two priests, Ernst Alt and Arnold Renz, grappled with her demons and recorded forty-two of the sessions on tape.

Anneliese Michel

Some half-dozen or more demons within Anneliese spoke, and even identified themselves. The Biblical figures Lucifer, Cain, and Judas were there, as were the historical figures Emperor Nero, Adolf Hitler, and others. Daily she did hundreds of genuflections, dropping to her knees until the ligaments were permanently debilitated. She had open sores all over her body. She scratched herself and bled. Her mouth and nose were raw, her eyes deeply bruised, her hair shredded. She was unbathed and stank horribly. She urinated on the floor and licked it up. Always her voiced growled back at the tormenting priests.

She refused food and drink and became a scrawny, wild creature in her own home. Her own family was afraid of her. The thinner and smaller she got, the more like an animal she became.

And then, one morning, the house was silent. The demons were gone. Anneliese lay in her bed, dead. She weighed 31 kg, or 68 lbs. She was 23 years old, a ragged, crazed stick-figure caricature of what she had once been. The cause of death was starvation and dehydration.

To exorcise (from the Greek *exorkizein)* means to adjure, to make a formal command, which must be followed by oath of obedience. The exorcist thus commands the possessing entity to take an oath that they will leave the host body. The practice probably predates written history. Songs for charming demons away are recorded in the Book of Psalms and in the Dead Sea Scrolls dated from some two thousand years ago. Various forms of exorcism have been practiced in virtually all cultures for as long as we have history.

Today doctors can look at cases like Anneliese, and though we cannot make a reliable diagnosis without an examination, it seems clear that she suffered from a variety of conditions including dissociative identity disorder (formerly called multiple personality disorder). It's usually comorbid (found alongside) with other psychiatric conditions such as schizophrenia, from which Anneliese probably suffered as well; and chronic stress is indeed one potential cause of her epileptic seizures. Although the psychiatric profession of the day had not been able to cure her conditions, it probably controlled them to some degree; as she didn't die until she stopped all medical treatment.

Belief is a key component of perceived possession and exorcism. If all parties believe that the sufferer is possessed, going through the motions of an exorcism may indeed solve the problem in some cases. And this is a serious problem, because that simple fact can be used to defend the practice; which sometimes results in preventable death.

After Anneliese's death, some within the Catholic church made an almost scientific effort to reform church laws governing the use of exorcism. When an exorcist speaks imperatively to the demon, instead of to the patient (to say "I command thee, unclean spirit," or some such thing), it confirms the patient's belief that they are indeed possessed by a demon. This confirmation by an authority makes the psychological problem much worse. Aware of this complication, a commission of conscientious German theologians petitioned the Vatican in 1984 to ban this part of the ritual. It took 15 years for the Vatican to render a decision. When they finally did revise the exorcism formula in 1999 (the first time it had been reviewed since the 17th century), it still allowed for exorcists to directly address the alleged demon. Thus, the Catholic exorcism rite remains contemptuous of basic ethics and any pretense of considering the patient's welfare to be important.

The commission had the additional motivation for reform when Anneliese's parents and both Alt and Renz were charged with negligent manslaughter for failing to call a doctor. During her final days, Anneliese's internal organ shutdown was probably irreversible, but a week before she could have been saved by even the simplest medical care. Recognizing that her parents had tried for many years to give every possible type of care for her episodes, the prosecution asked only for a fine for the priests and a guilty verdict but no punishment for the long suffering parents.

Before the trial, the parents had their daughter's body exhumed, based on a tip from a nun who claimed that she saw Anneliese's body was incorrupt in a vision. If true, it would confirm the supernatural nature of the case, proving that the exorcism was indeed the proper course of action. When the casket was opened, she was found to be decomposed as expected. The court found all four defendants guilty, but went further than the prosecution asked and gave suspended prison sentences plus three years probation. The Michels remained convinced that they'd done the right thing.

Of course, Anneliese was not the only victim of exorcism. The excellent website WhatsTheHarm.net lists over a thousand such cases, most from the recent decade, and most from Western countries. It is not ancient history and it is not limited to developing countries. Hundreds of professional exorcists walk among us, today, seeking critically ill psychiatric patients upon whom they can shout charms and sprinkle water. Many of these cases recount shocking tortures. Drownings, crucifixions, burnings, stabbings, all in the name of exorcism, and most to innocent children or the mentally ill.

When *The Exorcist* came into theaters, just as Anneliese was entering the worst of her final years, "possessed" people joined werewolves and zombies as favorite cinema monsters. It seemed that neither audiences nor filmmakers saw the patients as the victims, but as scary new antagonists to be feared. This perception has almost certainly hampered the efforts of those who want exorcism banned.

The exorcism dramatized in the movie was based on a real case. An anonymous young boy, given the pseudonyms Robbie Mannheim or Roland Doe, was exorcised by three priests in 1949. He survived, and went onto to have a successful family and career, but this is largely because his condition and exorcism were far less dramatic than the fictionalized version presented in the book and subsequent movie. One of the priests, Rev. Walter Halloran, gave a 1999 interview to *Strange Magazine* in which he revealed never having witnessed the bizarre incidents attributed to the boy: speaking Latin in a strange voice, having extraordinary strength, vomiting, markings appearing spontaneously on his skin. He did say the boy spat a lot and he saw the bed move once, but only when he leaned against it (it was on wheels). Nevertheless, the movie character based on this boy is one of cinema history's all-time most infamous monsters.

Anneliese had been a pretty young woman with striking black hair. As a child, she often played at her father's sawmill, and by all accounts her childhood was largely normal and happy. Even through the early years of her seizures, she was studying to become an elementary school teacher. Anneliese might be teaching in a Bavarian schoolhouse today, had she not been one of the unlucky few who were unable to make it through a troubled youth. She was a complex and talented person, who had humor and love and flaws and dreams.

I've given an example in this chapter of exploiting her torture and manslaughter to make a scary story. Filmmakers have exploited these victims to make not just *The Exorcist*, but a slew of other copycat films based on specific individuals, including Anneliese. Every time Linda Blair's head spun around, or she spat green vomit, we laughed and had a riotous old time at the theater. Would the same movies have been made exploiting the victims of other true-life crimes, and would we have laughed at the depictions of those actual victims in their dramatized death throes? For some reason, exorcism seems to have been given a pass, on the mistaken presumption that it is the victim who is the monster. These victims are often critically ill individuals — they may have medical or psychiatric problems that need treatment — they deserve neither to be tortured, killed via negligent manslaughter, nor to have their ordeal glorified as some kind of pop-culture horror story.

Exorcism is a brutal, heinous, medieval torture ritual justified only by ignorance. Its roots as a religious rite are irrelevant; a crime is still a crime. In this century, we have the means to actually help sick people. Do not condone the primitive obscenity that is exorcism.

References & Further Reading

Day, E. "'God told us to exorcise my daughter's demons. I don't regret her death'." *The Telegraph.* 27 Nov. 2005, Newspaper.

Editors. "The Exorcism of Anneliese Michel." *Online Demon Encyclopedia.* Demonicpedia, 7 Feb. 2010. Web. 6 Mar. 2011. <http://www.demonicpedia.com/tag/anneliese/>

Farley, T. "What's the Harm in Exorcisms?" *What's The Harm?* WhatsTheHarm.net, 7 Jan. 2008. Web. 6 Mar. 2011. <http://whatstheharm.net/exorcisms.html>

Goodman, F. *The Exorcism of Anneliese Michel.* Garden City: Doubleday, 1981.

Hansen, E. "What in God's Name?" *The Washington Post.* 4 Sep. 2005, Newspaper.

Opsasnick, M. "The Haunted Boy of Cottage City." *Strange Magazine.* 1 Jan. 1999, Number 20.

4. THE VOYNICH MANUSCRIPT

The true history and meaning of history's most famous undeciphered book.

Today we're going to look at the most famous undeciphered text of all time: A medieval book of science, full of beautiful illustrations and strange wisdom, and containing not a single word that anyone's been able to make heads or tails of: The Voynich manuscript.

So let's get the big question out of the way right up front. The Voynich manuscript is an unsolved mystery, at least so far. According to the best information we have now, we still don't know who wrote it, what it says, or what its purpose was. We do have some theories, but there will be no unveiling of a glorious answer today. However, the voyage of scientific exploration is always a fascinating one, and much of what we have learned is just as interesting as what we haven't.

Somewhere in Europe, probably northern Italy, sometime in the early 1400s, animals were slaughtered (either sheep, calves, or goats) and their skin turned into parchment. Probably very soon thereafter, someone, most likely two people, took a quill pen in hand and wrote a 38,000-word book using common ink, beginning to end, using an alphabet and language that have defied all identification. It's not a huge book, measuring about 16 by 23 cm (about 6 by 9 in), and about 5 cm (2 in) thick. There about 240 pages, most of them illustrated, the exact number depending on how you count pages that fold out into large diagrams or drawings, of which there are several. The alphabet has between 23 and 40 distinct characters, depending on how you classify some which may be decorative versions of others or two-character combinations.

The book has six sections, delineated by the types of illustrations. Section 1 is the largest at 130 pages, and contains detailed drawings of 113 plants and flowers that nobody has been able to identify. It's called the Botanical section. Section 2 is 26 pages of Astrological drawings; lots of circular and concentric diagrams, and some signs of the zodiac. The third section is called the Biological section and contains mainly drawings of nude women frolicking in intricately plumbed pools. Section 4 is the Cosmological section, featuring the most impressive foldouts that appear to be circular diagrams of some cosmic nature. The fifth section is Pharmaceutical, with over 100 drawings of herbs, roots, powders, tinctures, and potions of undecipherable contents or use. The last section, called Stars, is the most mysterious; it's 23 solid pages of text only, in short paragraphs, each marked with a star. Some of the illustrations show Eastern influence, including a probable map of the circular city of Baghdad, the center of Eastern knowledge.

A few hundred years later (we don't know exactly when), a cover was added, but unfortunately it's blank. Also at some later date, the illustrations were colored, by someone less careful than the original artist.

The book was owned by the English astrologer John Dee during the 16th century, who wrote his own page numbers in the upper right corner of each leaf. Dee sold it to Emperor Rudolph II of Germany, with the understanding that it was the original work of Roger Bacon, a 13th century friar widely regarded as one of the fathers of the scientific method. From there the book passed to one or two other owners, who wrote their names in it, and at one point it was presented to the scholar Athanasius Kircher in Rome along with a signed letter from a Johannes Marcus Marci expressing a hope that Kircher could translate it, in 1666. Marci's letter is still preserved with the book. The manuscript's history becomes unclear at that point, until it was discovered by antique book dealer Wilfrid Voynich in 1912 at the Jesuit college at Villa Mondragone in Italy. Voynich brought it to the attention of the world. After several owners, the book was eventually donated to its current home, the Beinecke Library at Yale University, under its official name of MS 408.

Since its discovery, hypotheses have abounded as to what the Voynich manuscript means. Many believe it's written in a type of code, but all efforts to find decodable patterns have failed. Some believe it may be what's called a constructed language, which is a language that's deliberately planned and invented rather than naturally evolved. Some have speculated it's to be used with a Cardan Grille, a paper with holes in it that you lay on top of the page and read only the revealed letters. Perhaps the most popular theory is that it's a hoax, written at practically any time since the parchment was made, and for just about any purpose ranging from financial gain to scholarly fraud to someone's weekend lark.

Guesses on its authorship are just as plentiful. Roger Bacon remains the usual suspect, but this is based only on the presumption of most of its owners and is not supported by any evidence. Roger Bacon never wrote anything else in the Voynich language so far as we know. Moreover, he died in 1294, more than 100 years before the book was written.

We can be certain of that, because we do know when the parchment was made, a fact that neither Voynich nor his predecessors could have known. Carbon dating of the parchment was performed at the University of Arizona in 2011 by Dr. Greg Hodgins, and nailed it down to the early 1400s. Dating the ink, however, is not something that we have any good way to do. Most ink can't be radiocarbon dated because it doesn't necessarily contain any organic matter; and even when it does, we don't have the technology to reliably separate carbon in the ink from carbon in the parchment. We've found that the pigments used are consistent with what is known to have been used in those years, but it could also be consistent with expert modern fakery.

However, we can still make some strong educated guesses. Parchment was commonly washed and re-used (it's a good way for modern forgers to create a document that legitimately radiocarbon dates to an ancient time), but doing so leaves chemical footprints. We do know that the Voynich manuscript was the first application of ink to its parchment. And from history, we know that parchment was always in good demand; it would have been virtually inconceivable for perfectly good parchment to sit unused for decades or centuries waiting for someone to come along and make a forgery on it. Combined with the fact that we have no reason to doubt the history of the book's ownership as given in Marci's letter, we can be pretty confident that the book was written about the same time as the parchment was made.

So let's look at the book's other properties to see what we can learn.

Here's an important one. There are no corrections in the book. There are also no places where the text has been squeezed to fit onto the page. This would be highly improbable if it were an original manuscript; we would absolutely expect there to be such minor errors in a first edition. So how do we explain this? There are a number of possible explanations, but two of them are most likely.

The first is that the book is a copy, perhaps even of something written by Roger Bacon. If a scribe has an original to work from, he can see how many words there are and properly plan everything to fit onto the page. And if he copies carefully, there will be no corrections. The copy theory is also consistent with other characteristics, such as its appearance of having been written straight through by only one or two people. If it is a copy, this alone doesn't tell us much that's useful in trying to decipher it. But it does leave us wondering why anyone would go to the trouble of making a nice copy of a book that doesn't say anything.

The second theory to explain the book's neat appearance is perhaps more revealing. The text could be complete nonsense, made up as the scribe went. There would be no need for corrections. There would be no need to compress the writing as space ran out.

The "complete nonsense" theory has one thing working against it. If it is nonsense, it's very good nonsense. It's almost too good to expect of an amateur. Computational analysis of the text has been run, exhaustively, many times by many different researchers, using many different techniques. This allows us not only to try and translate it (at which all attempts have met utter failure), but also to compare its metrics to those of actual languages. The letter frequency, word length, and word frequency are very similar to what we see in real languages. But they don't quite match those of any real languages. It's speculation, but I can imagine a monk or professional scribe who does this all the time being well aware of such things and deliberately giving the book a realistic appearance, but it seems less likely that an amateur, just a Joe Blow or professional from a different field, would happen to write gibberish that's such good gibberish.

But clues indicating that there is meaning within the text don't end there. Patterns of word usage and word relationships are also found within the six different sections, as if the sections are actually about different subjects. The pages within each section are more similar to each other than they are to pages in other sections.

However, a broader analysis of this leads to another interesting point, which we in the brotherhood often describe as "The plot thickens."

A famous analysis done in the 1970's by US Navy cryptographer Prescott Currier found that the Voynich manuscript is written in two distinct languages. He used the term languages, but also cautioned that they're also consistent with different subject matter, different encryption schemes, or possibly just different dialects. He called them Voynich-A and Voynich-B. Interestingly, Voynich-A and Voynich-B are in two different handwritings, though both use the same alphabet and script. Every page of the book is written entirely in either A or B. The Biology and Star sections of the book are written in Voynich-B; the others are written in Voynich-A. The exception is the first and largest section, Botanical, which contains some of each. But they're not simply interspersed. The way the book is bound uses bifolios, which are groups of pages folded together, which are then stacked on top of one another to be bound. Each bifolio in the Voynich manuscript is written entirely in one language or the other.

Actual sample of text

So let's wrap this up with my favorite theory, and the one that is perhaps best supported by all that's known. In the early 1400s, some professional, perhaps a physician or astrologer or alchemist, wanted to create some marketing material that demonstrated he had rare knowledge from the East. He engaged a monk or other scribe to produce a book filled with wondrous and curious illustrations from multiple sciences, and a text that nobody could read, which he could tell his customers was the source of whatever great Eastern wisdom he wanted. The monk had a colleague assist, and the two devised an alphabet and used their own multilingual familiarity with written language to devise a convincing nonsense text. It was well done enough that its owner could even use it to impress his colleagues. Thus, this anonymous professional ended up with impressive marketing collateral that's conceptually identical to the lab coat worn by a naturopath, the energy diagram on the wall of a yoga guru, and the purchased-online title of "doctor" sported by alternative practitioners of every variety.

This remains the leading theory. Not quite a hoax, and very deliberately and carefully created; yet full of nothing but pure nonsense. Perhaps one day the Voynich manuscript will reveal a different purpose, but for now, this theory is as good as any.

REFERENCES & FURTHER READING

Curators. "MS 408 Cipher Manuscript." *Beinecke Rare Book and Manuscript Library, Medieval and Renaissance Manuscripts.* Yale University, 10 Jun. 2002. Web. 29 Mar. 2011. <http://brbl-net.library.yale.edu/pre1600ms/docs/pre1600.ms408.htm>

Kennedy, G., Churchill, R. *The Voynich Manuscript: The Mysterious Code That Has Defied Interpretation for Centuries.* London: Orion, 2004.

Knight, K. *The Voynich Manuscript.* Los Angeles: University of Southern California, 2009.

Rugg, G. "The Mystery of the Voynich Manuscript." *Scientific American.* 21 Jun. 2004, July 2004.

Stolte, D. "UA Experts Determine Age of Book 'Nobody Can Read'." *UA News.* University of Arizona, 9 Feb. 2011. Web. 24 Mar. 2011. <http://uanews.org/node/37825>

Zandbergen, R. "The Voynich Manuscript." *The Voynich Manuscript.* René Zandbergen, 26 Feb. 2011. Web. 28 Mar. 2011. <http://www.voynich.nu/>

5. THE PORT ARTHUR MASSACRE

*Was a 1996 mass murder in Tasmania secretly a plot
by the government to get firearms banned?*

Today we're going to point the skeptical eye at a modern conspiracy theory, one that still stings sharply in the recent memories of Australians. In 1996, a young man from Hobart, 28-year-old Martin Bryant, loaded his car with guns and ammunition and went to the Port Arthur Historic Site, an old prison colony at the south end of the island of Tasmania. He killed 35 people and wounded 21 others, and was taken into custody after an overnight standoff and is currently serving a 1,035 year prison sentence without the possibility of parole.

For some, the Port Arthur massacre has become something like Australia's version of the Kennedy assassination. Some believe that no one man could have accomplished so much bloodshed. Some point to what they believe are discrepancies in the timelines of when Bryant is alleged to have been at various locations. Some believe the killings were too expert to have been done by anyone other than a trained killer. And finally, some believe that it was an elaborate conspiracy staged by the antigun lobby to provoke public sentiment. Many of the believers consider this to be evidence that the government wanted to terrorize the citizenry into banning all firearms, so that they might be able to exercise unchallenged tyranny. Such conspiracy theorists describe the event as a "PSYOP", a psychological operation by the government.

It should be fairly noted that few Australians believe any of the conspiracy theories, and a much larger number are angered by them, none more so than those who were there that day and escaped with their lives or watched their loved ones killed. But to evaluate the validity of a conspiracy theory, we take emotion out of the equation, and instead look to where the evidence leads.

Here is a quick summary of "the official version" of what happened. Martin Bryant, 28 years old with a clean-shaven baby face and long blond hair, had an estimated IQ of 66. He received a large inheritance from a friend, with which his family tried to purchase a bed & breakfast cottage called Seascape in Port Arthur, Tasmania. But another couple, the Martins, bought it first, which Bryant took personally. On April 28, 1996, he loaded his yellow Volvo with guns and ammunition, went to Seascape, and murdered the Martins. He then drove to the nearby Port Arthur Historic Site and went to the tiny Broad Arrow Cafe, where he ate lunch, then pulled out an assault rifle and in 15 seconds, killed 12 people and injured 10. In the next 90 seconds, he went into the nearby gift shop and killed 8 more people, most of whom were crouching to hide or trapped in the small room. He then moved to the parking lot, where he killed more people trapped between or on board parked buses.

This whole time, Bryant repeatedly fired at people who were running or hiding, but having no marksmanship skills, he missed everyone except those to whom he was able to get very close.

He got into his car and drove away, passing fleeing people, and stopped when he saw a young mother running with her two children. He killed all three at point-blank range. Finding the park exit blocked with cars driven by confused people unsure what was happening, Bryant went to a BMW, killed all four people inside it, transferred some of his guns and ammunition to it and drove away. He stopped at a service station where he killed a girl and forced her boyfriend, Glenn Pears, into the trunk of the BMW, and drove off again.

He returned to the Seascape cottage where he had begun his day, and fired at passing cars, injuring several more people. He took Pears inside the house, set the BMW on fire, and barricaded himself in. Police began to arrive and Bryant held them all off with gunfire. An 18 hour standoff lasted until the next morning, when Bryant killed Pears and lit the house on fire. He eventually ran outside, on fire, and was apprehended as he pulled off his burning clothes. In all, he'd killed 35 and wounded 21.

In the aftermath, new gun control laws were enacted throughout a shocked Australia. More than anything else, this is what sparked the speculation that a hidden government agenda must have motivated the entire episode, part of a giant master plan to trick the unsuspecting public into willingly disarming.

Much is made by the conspiracy theorists of the claim that Bryant was sent to prison for life without a trial, which would indeed be shocking and seemingly unprecedented. It's also misleading. Bryant pleaded guilty to all charges, so it didn't go to trial, like every case in which the defendant pleads guilty to all charges. Despite being of acknowledged low intelligence, he was found competent to stand trial, a finding that has not been challenged. His lawyer persuaded him to plead guilty simply because the evidence against him was overwhelming; he had no realistic chance of getting off, and a guilty plea was the route to the best possible outcome for him. It was not a conspiracy against a patsy; it was his best legal option.

That Bryant was placed in solitary confinement for the first eight months of his sentence is said to be evidence that the government didn't want him to be able to reveal any truths about the conspiracy. It's possible this is the reason, but there are at least two other reasons that Bryant, and many other criminals like him, are kept in isolation. The first is that among his victims were children, murdered at close range for no reason. Prison inmates have a reputation for not taking kindly to child killers, especially to those who need to use a gun to do it, and it's more than likely that Bryant would have been attacked or even killed in prison if not kept separated. Indeed, there were specific threats against him. Even his meals were prepared separately by special staff to prevent anyone from trying to poison him. The second reason is that he was on suicide watch and was in a special hospital ward suicide-proof cell, and for good reason; he's attempted suicide at least twice so far.

Some conspiracy theorists claim that Bryant displayed extraordinary combat skills that could only belong to a highly trained expert, and not to an intellectually challenged kid with no firearms experience. One noted that the true perpetrator must be one of the top 10 or 20 shooters in the entire world. In fact Bryant displayed no special skills, killing nearly all of his victims within just a few meters, and some with the muzzle of his gun actually touching them. He missed all of his shots that were at any appreciable distance. Nor should it be surprising that the Port Arthur killer would be untrained; it's quite common for mass killings to be carried out by loners with no military connections or special training.

And then there are myriad small details on which some sources are unclear; for example, whether the knife with which Bryant killed Mr. Martin at Seascape was found in Bryant's bag in the Broad Arrow Cafe, or nearby. Some characterize discrepancies such as this as evidence that the knife must have been planted by police. There is also minimal publicly known evidence that physically places Bryant at the Port Arthur Historic Site at all on that day. There is speculation surrounding the appearance of an armed man on the roof of a building at Seascape cottage during the night. One need only scan through any of the many web sites promoting the idea that the Port Arthur massacre was a government conspiracy to find many such questions raised.

But there is an alternate explanation for all of these questions that satisfies the available evidence without the need to introduce a conspiracy. Whatever evidence might exist is evidence in a murder case. It is not necessarily available to the public. Whatever it was, it was described by Bryant's attorney as overwhelming, and was sufficient for the prosecutors to charge him. Bryant was caught red-handed during the siege; there is no plausible doubt that Martin Bryant is the person who held off the police overnight. Whatever physical evidence may have been gathered by investigators that supports the chain of eyewitness accounts, all the way back to Martin Bryant's yellow Volvo laden with weapons and recovered at the Port Arthur parking lot, is sealed. This evidence's apparent nonexistence may indeed be consistent with a cover-up, but it's also exactly what we'd expect to find in the context of a murder investigation.

Once, Bryant's attorney took some photos of him in jail during a visit, which were then confiscated and destroyed by prison authorities. This incident is often pointed to as proof that some kind of cover-up is taking place, along with assertions that nobody has ever been allowed to photograph Bryant; perhaps because it might be discovered that his physical description does not match that given by eyewitnesses. This is a goofy claim. Photos and video of Martin Bryant were widely published throughout the media following the incident, and are still all over the Internet to this day. Does it really make sense that a conspiring Australian government would think it was accomplishing anything by banning photographs of Bryant? The lawyer's photographs were destroyed because cameras are not permitted inside prisons without prior permission, for obvious reasons, and he had failed to request any such permission. Again, there is no conspiracy needed to explain these events.

So why do the conspiracy theories persist? Why are some people so quick to jump on board any bandwagon that presumes the existence of a hidden malevolent power? It's yet another manifestation of the way our brains are hardwired. We want to find patterns. We want to make connections between cause and effect. When something goes bump in the night and nothing is seen, our brains want to assign the blame to a ghost. When shapes in a photograph from Mars mimic a face, our brains conjure up a Martian civilization that must have carved it. When a natural disaster happens, we look to secret government research as the culprit. And when a lone gunman murders 35 people, it's natural for our brains to imagine an evil intelligence behind what happened. Psychologists call this agency detection. The caveman who errs on the side of caution and assumes that every rustle in the grass is a saber-toothed cat is more likely to survive than the one who casually dismisses it as a harmless breath of wind.

As aggravating (or even offensive) as they might be, conspiracy theories like the Port Arthur massacre are the naturally evolved result of our brains failing on the side of caution. In this case it's wrong, and in many other cases too. But if we always assume that the Martin Bryants of this world are troubled loners acting completely on their own, evolutionary theory says that one day the saber-tooth will get us. We have to always look at the facts.

REFERENCES & FURTHER READING

Altmann, C. *After Port Arthur.* Crows Nest NSW: Allen & Unwin, 2006. 9-23.

Angle, M. "Port Arthur conspiracy theory still upsets Tasmanians." *The World Today.* Australian Broadcasting Corporation, 22 Feb. 2001. Web. 25 Mar. 2011.
<http://www.abc.net.au/worldtoday/stories/s250296.htm>

Bingham, M. *Suddenly One Sunday.* Sydney: Harper Collins, 1996.

Editors. "Shedding Light on Port Arthur Killer." *The Age.* The Age Company Ltd., 29 Mar. 2006. Web. 31 Mar. 2011.
<http://www.theage.com.au/articles/2006/03/28/1143441154819.html>

Mullen, P. *Psychological Report, Martin Bryant.* Melbourne: Victorian Forensic Psychiatry Services, 1996.

Stein, G. "Managing Martin: The Jailing of Martin Bryant." *Background Briefing.* ABC Radio National, 16 Mar. 1997. Web. 31 Mar. 2011.
<http://www.abc.net.au/rn/talks/bbing/stories/s10603.htm>

6. FINDING THE POW/MIAS

Are American POW/MIAs still being held captive inside Vietnam?

It was 1985 when John Rambo peered through the jungle greenery and saw a bamboo cage full of aging Americans, dirty, sweaty, and bearing fresh wounds from their daily beating. Vietnamese guards paced the compound, as they had for the past 15 years, despite the war having ended over a decade earlier, and there remained little useful intelligence to be gained from interrogation. It  was Hollywood's envisioning of the rumor that American POW/MIAs (Prisoner of War, Missing in Action) are still held captive somewhere. To some, it's a way of holding out hope that a loved one is still alive and may even make it home someday. To others, it's another conspiratorial evil of the American government, which is alleged to know that the prisoners are out there but refuses to acknowledge them or make any effort to bring them home. Are there POW/MIAs still out there somewhere? What does the military actually know?

Soldiers remain unaccounted for from every American war, but those most associated with the POW/MIA movement are the Vietnam servicemen. They are the ones Rambo was searching for, and they're the ones we're going to try and find today.

So let's start by defining exactly who we're looking for. The official numbers given are taken as of Operation Homecoming, a diplomatically negotiated prisoner exchange at the start of 1973. At this time, all known American POWs were released, totaling 591 men. It is the conclusion of the official agency, the DPMO (Defense POW/Missing Personnel Office), that that 591 does represent all prisoners taken, with the exception of 113 who died in captivity. Those soldiers remaining unaccounted for at that time comprise the subjects of the POW/MIA issue.

That remaining total was, at the time, 2,646, all categorized as missing in action; none considered to still be alive as prisoners of war. However, there are several other published numbers that are higher. The reason is that these numbers include some combination of missing civilians, other foreign nationals working with the Americans during the war, casualties from other actions in Vietnam through 1975, and servicemen who are known to have been KIA (killed in action) but their bodies were not recoverable for whatever reason. The smaller 2,646 number excludes KIAs.

As of April 2011, 953 of these have been accounted for. These accountings have included remains that have since been recovered and identified, remains that were recovered but were not identified, and there are even a few men who returned to the United States alive on their own, mostly in the 1970s. This leaves 1,693 Americans who are MIA and unaccounted for, at least officially; but the documentation supporting that number is pretty strong. Even if that number is off by even as much as a few hundred one way or the other, it doesn't have much bearing on the important question. So please don't email me that you have a different number; although individual guys are important, it's really hard to get totals upon which everyone agrees. Different groups publish different numbers. They're all in roughly the same ballpark.

It's worth noting how much better documented the numbers are than in previous wars. The unaccounted for soldiers in

World War II and the Korean War totaled 14% of all casualties; in Vietnam, the 1,693 number comes to only 3%. But really, that's neither here nor there; what we want to know is whether any of those 3% are still held in captivity somewhere, and whether the US government knows about it and is covering it up.

There are a few obvious possible fates for many of those men. One is that they were in fact killed in action, and their bodies were buried by the enemy in graves that will probably never be found. Undoubtedly this is what happened to some of the MIAs; how many, we'll probably never know.

Another possibility is that they deserted. Stories of some deserters are known. 40 men are known to have deserted, which may sound like a dubiously small number. Some returned to the United States under false identities; some escaped to other countries such as Australia or Canada; and some found homes inside Vietnam, both in remote villages and in the larger cities such as Saigon (now called Ho Chi Minh City).

One such home was the legendary "Soul City", an apocryphal district in a Saigon suburb populated by African-American servicemen, most of whom were presumably deserters. From this stronghold, they are said to have engaged in drug dealing and black marketeering. Soul City was never officially acknowledged or busted, though its existence was something of an open secret. Could some of the MIAs simply be soldiers who stayed at Soul City? It's certainly possible.

However, any deserters who hoped to carve out a life for themselves inside Vietnam would have had a pretty ugly time at the fall of Saigon on April 30, 1975. North Vietnamese forces swept Saigon and other cities throughout the country. No American prisoners are known to have been taken during this operation. No doubt the North Vietnamese did encounter some American deserters. We'll never know how many or who they might have been, but it's probable that they were killed. North Vietnamese would have assumed that any leftover

Americans they ran into were CIA. Any enclaves like Soul City would have been overrun and destroyed.

This leaves the other frequently suggested fate of some of the 1,693 MIAs: That they are still held prisoner as POWs somewhere, and the United States government knows it, and is covering it up. If this conspiracy has indeed been discovered, that means someone somewhere must have discovered it and knows where these alleged prisoners are.

Rambo's bamboo cages are not the only place such POWs might be held. In 1992, Russian president Boris Yeltsin issued a statement that some American POWs had been transferred from Vietnam to prison camps inside the Soviet Union, where he said it was "very possible" they may still be alive today. This revelation sent shockwaves throughout the POW/MIA community. President George H.W. Bush and the bipartisan leaders of a Senate subcommittee investigating the POW/MIAs praised the announcement, and praised the vow of a new joint US-Russia POW/MIA commission to redouble its efforts.

Yeltsin is not the only highly-placed official to have made such comments. For a year and a half in the early 1990s, a Senate Select Committee on POW/MIA Affairs was formed that included both John Kerry and John McCain, mainly to investigate charges of conspiracy and cover-up by Secretary of Defense Dick Cheney and the current Bush administration. Two former Secretaries of Defense, Melvin Laird (January 1969 - January 1973) and James Schlesinger (July 1973 - November 1975), testified to this committee that they believed there were still prisoners left behind. To many, that sounds like about the most authoritative statement possible and proves the case. So how can it be that the committee eventually concluded that no compelling evidence exists that any Americans remained alive in captivity?

Secretaries of Defense know only what their staffs tell them. Laird and Schlesinger had served during and immediately after the war, at a time when none of the many reports had yet been thoroughly investigated, and at a time when American investigators did not have access to the battlefields and incident sites. What Laird and Schlesinger testified to was what their staffs knew in the early 1970s. But ever since the late 1980s, the US has had full access to those sites, including cooperation from Southeast Asian officials, and has maintained research offices on the ground throughout the region. By the time of the Senate Select Committee, we had a much fuller and clearer picture of what had actually happened at the close of the war.

Where Yeltsin got the information backing his claim is not known, but it is known that no reliable evidence was ever produced that supported it. Some have speculated that Yeltsin simply asked his staff for something he could announce to give the appearance of *glasnost* and increased cooperation between the two great superpowers. No prisoners from any American wars have ever been found held captive in Russia, China, Korea, or anywhere else. The DPMO has more such reports than you can shake a stick at, but all that have been considered reliable enough to investigate have led nowhere.

And it's this lack of any living POWs rescued that has fueled not only the conspiracy theory that the government is covering up their existence, but also a protracted and ugly battle of words between those within the DPMO and a massive activist community spearheaded by veterans who collect their own reports, and remain firmly convinced that prisoners are still being held. Twenty years ago many of these activists were still in the field, doing their own searches throughout Southeast Asia; these days the community has aged and it's a bunch of grouchy old men calling each other liars on web sites. Most of them appear to me to be split along political party lines, with conservatives generally supporting the claims of a cover-up, and liberals generally supporting the findings of the Department of Defense. Tribalism at its finest. I have no doubt that much of the feedback I receive will criticize me for not including the astonishing discoveries of researcher so-and-so; there are too many of them to list.

The Library of Congress maintains a web site, POW/MIA Databases and Documents, with over 150,000 declassified documents from the DPMO that are fully searchable by name. Search on the name of an MIA or POW and a list of documents will come up. Many of these are quite interesting, giving reports of sightings or descriptions of photographic analysis. Some consider this database to be an example of the transparency of the DPMO's efforts; others consider it fool's gold made to pacify the gullible.

Either way, claims that the government is ignoring the issue are hard to defend. More of the 1,693 are identified and put to rest each year, as can be seen on the DPMO's web site. In 2001, all sixteen members of a joint American-Vietnamese team searching for American MIAs were killed when their helicopter crashed in Vietnam on their way to excavate a site searching for remains to identify. Ironically, another such team was diverted from Laos to identify them. The war is over for many, but still very much a daily danger to some.

Some of the 1,693 probably died in combat. Some of them probably deserted. Some are probably living under assumed identities anywhere in the world, and many of those have likely died of natural causes. Are any of them held prisoner? We can't say that they're not, but we can say the evidence has not yet been convincing enough. Volumes of shoddy evidence don't aggregate into good evidence. You can stack cow pies as high as you want; they won't turn into a bar of gold.

REFERENCES & FURTHER READING

Burkett, B., Whitley, G. *Stolen Valor: How the Vietnam Generation Was Robbed of Its Heroes and Its History.* Dallas: Verity Press, 1998.

DPMO. "Vietnam Era Unaccounted For Statistical Reports." *Defense Technical Information Center.* US Department of Defense, 1 Apr. 2011. Web. 16 Apr. 2011.
<http://www.dtic.mil/dpmo/vietnam/statistics/>

Eaton, W. "Nixon Defense Secretaries Say U.S. Left POWs in Vietnam." *Los Angeles Times.* 22 Sep. 1992, Newspaper.

Federal Research Division. "The Vietnam Era POW/MIA Database." *POW/MIA Databases & Documents.* Library of Congress, 17 Jan. 1999. Web. 16 Apr. 2011. <http://lcweb2.loc.gov/frd/pow/powhome.html>

Hendon, B., Stewart, E. *An Enormous Crime: The Definitive Account of American POWs Abandoned in Southeast Asia.* New York: Thomas Dunne Books, 2007.

Schlatter, J. "MIA Facts." *MIA Facts Site.* Col. Joe Schlatter, 25 Jan. 1999. Web. 16 Apr. 2011. <http://www.miafacts.org/>

Senate Select Committee. *Report of the Select Committee on POW/MIA Affairs.* Washington DC: United States Senate, 1993.

7. SUPERHUMAN STRENGTH DURING A CRISIS

Popular stories tell of mothers lifting cars off their children. Can the human body really do such feats?

Today we're going to recount heroic tales of superhuman feats of strength, when in the face of disaster, some people are said to have summoned up incredible physical power to lift a car off of an accident victim, move giant rocks, or like Big John of song, single-handedly hold up a collapsing beam to let the other miners escape. Are such stories true? There are many anecdotes supporting the idea, but we're going to take a fact-based look at whether or not it truly is possible for an adrenalin-charged person to temporarily gain massive strength.

In proper terminology, such a temporary boost of physical power would be called hysterical strength. The stories are almost always in the form of one person lifting a car off of another. In one case in Colorado in 1995, a police officer arrived at a single-car accident where a Chevy Chevette ended up on top of a baby girl and sank into the mud. The officer lifted the car and the mother pulled the girl out. In 2009, a man in Kansas lifted a Mercury sedan off of a six-year-old girl who had been trapped underneath when it backed out on top of her. In 1960, a Florida mom lifted a Chevy Impala so that a neighbor could pull out her son, who had become trapped when he was working on the car and his jack collapsed. There's even the case where the MD 500D helicopter from *Magnum, P.I.* crashed in 1988, pinning the pilot under shallow water; and his burly friend (nicknamed Tiny) ran over and lifted the one-ton helicopter enough for the pilot to be pulled out. And, of course, the list goes on, and on, and on.

In each of these cases, some aspect of leverage or buoyancy probably played some role in reducing the magnitude of the feat to something more believable. And even lifting many cars by several inches still leaves most of its weight supported by the suspension springs. But our purpose today is not to "debunk" any of the specific stories. The majority of them are anecdotal, and interestingly not repeatable; in many cases, the person who summoned the super-strength later tried it again only to find that they couldn't do it. Basically, what we have is a respectably large body of anecdotal evidence that suggests that in times of crisis, danger, or fear, some people have the ability to temporarily exercise superhuman strength.

The typical explanation given centers on adrenalin. Adrenalin, also called epinephrine, figures prominently in what's popularly called the "fight or flight" response. Sudden stress, such as an impending fight or other dangerous situation, triggers the sympathetic nervous system to induce the fight or flight response, sometimes called hyperarousal or the acute stress response. It's a way that your body readies itself to deal with physical harm, very much like calling "Battle stations!" on a warship. The adrenal gland releases adrenalin into your bloodstream, and as it spreads throughout your body, it does different things to different types of tissue. Your airways relax to maximize breathing capacity, and metabolism increases. Your muscles go into glycolysis, which produces energy-rich molecules fueling them for extraordinary action. While blood flow to the muscles is increased, blood flow to vulnerable extremities is decreased. Dopamine is produced in the brain as a natural pain killer. Peripheral vision turns into tunnel vision to minimize distractions. Reflexes and reaction times improve. Non-critical functions like digestion and sexual function slow or even stop.

All of this is physiological fact. But how much does it really do for you? Is it the marginal improvement in ability that it would seem to be, or can it really supersize your strength by a huge factor of three, five, or ten? Fortunately, this has been studied.

Let's look at one such case, when a Camaro ran over a bicyclist in Arizona in 2006. The superhero was Tom Boyle, who lifted one side of the more than 3,000 pound car so that the wheels were off the ground. What's interesting in this case is that Tom Boyle is a power lifter. He's a huge guy, and he knew exactly how much he could dead-lift: 700 pounds. The world record at the time was 1,008 pounds. Neither number would be enough to get two of the Camaro's wheels off the ground.

As detailed in Jeff Wise's book *Extreme Fear*, the puzzle of weightlifters such as Tom Boyle has been at least partially solved by Vladimir Zatsiorsky, a biomechanics specialist at Penn State, who has published his results in his book *Science and Practice of Strength Training*. Zatsiorsky gives three numbers to describe an athlete's lifting potential. The highest is your absolute strength, which is the theoretical maximum that your muscle fibers, tendons, and bones can take. This number can never be exceeded, and realistically, can never be quite reached. The lowest is your maximal strength, which is the maximum you could lift using conscious effort, in a gym or other controlled environment. According to Zatsiorsky's research, the maximal strength of most ordinary people is just about 2/3 of their absolute strength. This means that for a person who can lift 200 pounds, 300 pounds is their frame's theoretical maximum. Tissues would fail, preventing that person from lifting any more, no matter what fight or flight response came into play.

However, for trained weightlifters who practice lifting their maximum every day, this number is higher, about 80%. Taking Tom Boyle's 700 pound maximum lift as an example, we calculate that 875 pounds is the maximum that his body could have taken before structural failure.

But here is where Zatsiorsky found that things get interesting. Somewhere in between the maximal strength and the absolute strength is a middle ground that appears when the body goes into competitive mode. The fight or flight response also appears when faced with the pressure of competing. Some research has even found that simply shouting encouragement toward competitors can even physically improve performance. Zatsiorsky has measured some athletes reaching as high as 92% of their body's absolute strength during the most intense competitions. This explains why world records are often set at the Olympics; there is no venue that places more pressure on athletes.

If Tom Boyle had this much pressure on him, and it's more than likely that a young man's imminent death is more stressful than the possibility of winning a contest, he could have deadlifted perhaps 800 pounds. It's still not enough to get two of the car's wheels off the ground. No matter what inspiration, mental mode, or adrenalin rush came over him, lifting much more than that would have resulted in structural failure, and the car would have fallen and crushed the young cyclist.

Sadly for the anecdote, we must conclude that the story is not accurate as reported. Perhaps both wheels didn't actually leave the ground and the suspension was doing some of the work. Perhaps Boyle lifted from the lighter back end of the car and not the heavier side. Perhaps the car was inclined in such a way that the leverage angle was more favorable. Whenever these rare events happen to be captured on film, some such advantage always comes into play. When Tiny was said to have "lifted the helicopter", it's pretty clear that he merely rocked the bulbous craft as it was lying on its side against the sloped and uneven riverbank.

So when we shine the light of science upon the notion that a dangerous situation can breed superhuman strength, we find that it's only partly true. The fight or flight response can indeed help you exceed your normal capabilities, but only by some limited increment; not enough to justify the common perception of these extraordinary stories.

But fight or flight is not the only possible booster available. There are other similar stories that have nothing to do with lifting a car off an accident victim. One such example is that of criminals fleeing police, fighting people off with apparent invincibility, shrugging off attacks and making an escape. The illegal drug PCP is sometimes offered as an explanation, and it can indeed be a good one. Among PCP's many psychological effects are a *feeling* of great strength and power; and among its physical effects is the blocking of pain. Put these two together, and an idiot on PCP can do just about anything. He won't actually have any greater strength, but he'll think he does; and as a result, such clowns often end up with a myriad of self-inflicted injuries, sometimes even killing themselves.

Such drugs have also been suggested to explain groups such as the Norse berserkers, a subset of Viking shock troops who fought like enraged wild animals, impervious to pain, and contemptuous of injury. Some researchers have suggested that berserkers may have taken hallucinogenic mushrooms before going into battle, as did Zulu warriors. Another theory states that they may have simply gotten really drunk, but this likely would have resulted in poorer performance in battle. It's also possible that berserkers simply worked themselves up into a frenzy, and combined with the fight or flight response to the impending battle, did indeed gain heightened physical ability.

Anecdotal evidence has suggested that cases of electric shock have caused people to launch themselves across the room involuntarily, presumably because the electricity fires the muscles. If true, it would support the notion that the musculoskeletal structure is indeed capable of feats that exceed even Zatsiorsky's absolute strength measurements. But this is not only unproven (and obviously untested), it is an incomplete theory. If such a thing has happened, exhibiting strength that has in fact exceeded the absolute strength, there is no evidence that it was done without crippling injury to the muscles or tendons.

The ability to acquire superpowers, even if only temporarily, is such a compelling possibility that most of us really want these stories to be true. And many of them probably are true to some degree, just exaggerated, misreported, or even misinterpreted by those who were there; and so, sadly, they're not yet the confirmation of superpowers that we're hoping for. It's a really intriguing field of research, and an attractive goal. But it's a goal we'll only reach if we go beyond the popularly reported versions of the stories and take the trouble to learn what's really going on.

REFERENCES & FURTHER READING

Campenella, B., Mattacola, C., Kimura, I. "Effect of visual feedback and verbal encouragement on concentric quadriceps and hamstrings peak torque of males and females." *Isokinetics and Exercise Science.* 1 Jan. 2000, Volume 8: 1-6.

Fabing, H. "On Going Berserk: A Neurochemical Inquiry." *Scientific Monthly.* 1 Nov. 1956, Volume 83: 232.

Foote, P., Wilson, D. *The Viking Achievement: The Society and Culture of Early Medieval Scandinavia.* New York: St. Martin's Press, 1970. 285.

Friedman, H., Silver, R. *Foundations of Health Psychology.* New York: Oxford University Press, 2007.

Walker, A. "The Strength of Great Apes and the Speed of Humans." *Current Anthropology.* 1 Apr. 2009, Volume 50, Number 2: 229-234.

Wise, J. *Extreme Fear: The Science of Your Mind in Danger.* New York: Palgrave Macmillan, 2009.

Zatsiorsky, V., Kraemer, W. *Science and Practice of Strength Training.* Champagne: Human Kinetics, 2006.

8. THE SECRET OF PLUM ISLAND

Does this secret government lab really create genetic mutants and biological weapons?

The very name of the place conjures up images from H.G. Wells or Dean Koontz, like strange hybrid creatures creeping through the island's greenery at night. Popular urban legends tell of carcasses of freakish unknown beasts washing ashore on nearby beaches. And the armed guards pacing the secure facility seem to confirm just about any nefarious rumor you might hear about Plum Island, a government research facility off the shore of New York that intrigues monster hunters, conspiracy theorists, and the merely curious alike.

When the corpse of the strange-looking "Montauk Monster" washed ashore in Montauk, NY in 2008, the media's first speculation was that it could have been a genetic experiment that somehow escaped from Plum Island. More recently, conspiracy theory promoter Jesse Ventura hyped up Plum Island on his television show as some kind of secret biological weapons research lab, that's not only illegally producing biological weapons, but that also poses an imminent threat to the region should any of these pathogenic compounds be released accidentally. Plum Island seems to have all we need to warrant a close examination with our skeptical eye. What do they actually do at Plum Island, and are there really strange mutant creatures being produced there?

Plum Island's association with the United States government goes all the way back to the 18th century, when General George Washington recognized its strategic importance. The island is located squarely in the narrow mouth of Long Island

Sound, which contains most of New York state's important harbors. It's home to the Plum Gut Lighthouse and Fort Terry, constructed in 1897, and used through World War I as an artillery base to protect New York from invading ships. Geographically the island is unimpressive; largely flat, a few low dunes and bluffs, a well wooded strip of island less than 5 kilometers long, and about 840 acres.

After World War II, Plum Island's handy combination of safe isolation from shore plus convenient access made it attractive to the US Army Chemical Corps, which planned and began construction of a facility there. However, those plans were ultimately canceled before construction was complete, and the new buildings were taken over in 1954 by the United States Department of Agriculture to study foot and mouth disease, at what became the Plum Island Animal Disease Center.

Foot and mouth disease is a highly contagious viral infection that can spread very quickly throughout animal populations. Hooved animals are mainly affected, most often cattle, but other animals can spread it as well. It's rarely fatal, but since it renders entire populations of livestock unsuitable for meat or milk production, it can rapidly pose a huge threat to the economy and food supply of a whole region or even a nation. Epi-

Plum Island

demics continue worldwide to this day, most recently in Bulgaria, Japan, Korea, and the United Kingdom, requiring the destruction of millions of pigs and cattle.

So with foot and mouth disease being a very real and imminent danger to the security of the United States, the USDA had good reason to set up the Plum Island facility. The work focused on developing vaccines and treatments for the disease, but since it could also be used as a biological weapon against the economy of an enemy nation, Plum Island researchers looked into this as well for a time. But in 1969, President Nixon stopped all non-defensive research into biological weapons, and shortly thereafter, the US ratified the Geneva Protocol and the Biological Weapons Convention, which outlawed biological warfare. Plum Island continued its work developing defenses against the disease.

And not just foot and mouth disease, but others as well. African swine fever, vesicular stomatitis, rinderpest (aka cattle plague) and other diseases also represent clear and present threats to food production, and the scientists who work at Plum Island (mostly microbiologists) are attacking these diseases too. Life inside is a bit like something out of science fiction. The entire facility is negatively pressurized, so that air flows in and never out, thus keeping any airborne germs from escaping. The center of the facility, where the live animals are kept, is at even lower pressure, keeping air flowing away from the scientists to protect them. Entering requires changing into autoclave-sterilized garments, and leaving requires a full body scrubdown with disinfectant through an airlock.

Since 9/11, security at Plum Island has been under increased scrutiny, as one theoretical threat would be for an enemy force to steal pathogens and infect our livestock population. So in 2003, the Department of Homeland Security took over administration of the island, while the Department of Agriculture still does the research. A somewhat infamous Government Accounting Office report released later that same year outlined many flaws in the security that needed to be addressed, producing a long list of recommendations, all of which have since been implemented. Plans are now in place that guard against the increased threats that Homeland Security perceives.

In the end, Plum Island may fall victim to more mundane forces. It's currently preparing to reduce staff in the face of government budget cuts; and the island has in fact been listed for sale since 2010, due to a 2008 decision by the Department of Homeland Security to move it to a more secure location in Kansas. According to the Department of Agriculture, technology is now sufficiently advanced that the study of foot and mouth disease can now be safely continued on the mainland, so little reason remains to continue the research on an inconvenient, outdated island facility.

Moreover, the safeguards at Plum Island have proven themselves inadequate in the past. Twice in 2004, foot and mouth disease was accidentally released inside the biocontainment area. In separate incidents, two head of cattle and four pigs were found to be infected even though they shouldn't have been. Both incidents took place inside the innermost, negatively pressurized animal pen areas, and so posed no outside threat. However, many years before, in 1978, animals kept in outdoor pens on Plum Island were found to have been infected with foot and mouth disease, and this was what prompted the current practice of keeping all research animals inside the biocontainment area.

There was also much fallout from a highly publicized strike in 2002, when maintenance workers employed by a private contractor on the island went on strike. It was the first time a strike had ever occurred against a secure government laboratory, and since Plum Island's safety cannot be maintained without maintenance work, the strikers were quickly fired and replacement workers were hastily brought in. Questions were rightly raised about the training and security clearances of the replacement workers, so this underscored yet another vulnerability at the facility. This too was addressed by the Homeland Security recommendations in 2003.

The claims of secret research or clandestine development of biological weapons are a bit naive. Much of the scientific staff consists of visiting scientists from other nations and quite a few from Yale University and the University of Connecticut. All of them even put their contact information on Plum Island's web site. Conspiracy theorists are free to contact anyone currently or formerly employed at Plum Island and will probably find an enthusiastic researcher eager to discuss his work. Evidence of secrecy or cover-ups has simply never surfaced, and would be pretty hard to hide in such an open environment. There's nothing classified or secret about any of the research done at Plum Island.

But urban legends continue to propagate. One author, Michael Carroll, wrote a 2004 book called *Lab 257: The Disturbing Story of the Government's Secret Plum Island Germ Laboratory* in which he asserts, among other things, that Plum Island created Lyme disease, and even caused a 1975 outbreak of it. Carroll also suggested that other outbreaks on the mainland were caused by Plum Island, including Dutch duck plague in 1967 and West Nile virus in 1999. Although his many fanciful claims (supported only by his own speculation) are not taken seriously by people who know better, the general public generally doesn't; and when the only publicized information about a place is created by people like Jesse Ventura and books like this, it should surprise no one that the rumors persist.

So how do we know that they're just rumors? We don't always, but we can get a pretty good idea. The publications from the scientists who work there and at the associated universities paint a clear enough picture of what does happen at Plum Island, but that's not proof that other nefarious activities don't also happen in secret. Many times, conspiracy theorists have accused me of being a paid disinformation agent for the government. Try proving that I'm not. You can see what I do with my time; you can interview my family and friends; but that won't prove that I don't have a secret room and carve out secret blocks of time when nobody's around.

You also can't prove that I *am* a disinformation agent. I'm not (I don't even know if there's any such thing), so obviously no evidence exists that I am. Those charges are based purely upon speculation. Similarly, claims about secret genetically designed creatures coming from Plum Island are based purely upon conjecture drawn from the air. We do know from history that their research included the development of biological weapons prior to 1969, but no evidence exists that it has continued; and there's never been anything to suggest that they ever genetically created monstrous hybrid animal species, nor any reason for them to do so. You can speculate all day, and say that this sounds plausible or that it would justify distrust of the government, but it doesn't make it so.

Just to give one example, Jesse Ventura's main source for his TV show about Plum Island was Kenneth King, author of *Germs Gone Wild,* and a former writing teacher and attorney who has no more inside knowledge about Plum Island than you or I. His book points out that accidents, such as those that happened at Plum Island in 2002, frequently occur at such labs throughout the country and all over the world. Of course this is true, but this is an entirely different question than whether Plum Island is secretly and illegally developing clandestine bioweapons. This was just the best that Jesse could do, and his technique of raising alarm about something that *appears* relevant, or suggesting guilt by association, is a common one among conspiracy theorists who have no valid evidence.

Next time you enjoy a steak or a pork chop, thank the Department of Agriculture for keeping foot and mouth disease in check. And, by the way, the Montauk Monster was just a raccoon.

References & Further Reading

Carroll, M. Lab 257: *The Disturbing Story of the Government's Secret Plum Island Germ Laboratory.* New York: Morrow, 2004.

Cella, A. "An Overview of Plum Island: History, Research, and Effects on Long Island." *Long Island Historical Journal.* 1 Sep. 2003, Volume 16, Number 1-2: 176-181.

GAO. *Combating Bioterrorism: Actions Needed to Improve Security at Plum Island Animal Disease Center.* Washington, DC: United States General Accounting Office, 2003.

Rather, J. "Plum Island Reports Disease Outbreak." *New York Times.* 22 Aug. 2004, Newspaper.

Schultz, E. "Homeland Security: Plum Island To Suspend Research And Development If Government Shuts Down." *New London Patch.* 8 Apr. 2011, Newspaper.

USDA. "An Island Fortress for Biosecurity." *Agricultural Research.* 4 Dec. 1995, Volume 43, Number 12.

9. Spontaneous Human Combustion

People can catch on fire... but can it really happen when there is no external source of ignition?

Today we're going to point our skeptical eye at one of the mainstays of the paranormal: spontaneous human combustion (SHC). The idea is that people can, while simply minding their own business, burst into flames, with no source of ignition. It's not a medically recognized phenomenon, and no explanation exists that can reasonably account for any but a few of the many stories. Thus, it's found a firm home in the world of the strange, that subdivision of Earthly phenomena that is studied and promoted by only a few fringe researchers and outsiders. That doesn't mean it's wrong though; and we're going to look at it as closely as we can.

Spontaneous human combustion is a little different from most paranormal phenomena, in that it's a claim of no external source for the fire. That people have burned up is the fact that's not in question; the question is the theory of what *caused* them to burn. In this case, believers are asserting that there was no conventional cause. Their job is, in effect, to prove a negative. Proving a negative is different from the null hypothesis. The null hypothesis for an unexplained fire is simply to say that no cause is known, which is different from stating authoritatively that there *is no* cause. Thus, the burden of proof still rests on the claimant, even in this case. Science does not allow us to make the jump from "the cause of the fire is unknown," therefore "the cause is known and it's spontaneous human combustion." Science allows unanswered questions; indeed, science exists because of unanswered questions. The lack of an answer proves only that we don't know something yet; it does not prove the existence of the paranormal.

Stories of SHC generally fall into one of two categories. The first type is the discovery of a body that burned while nobody was present, usually almost completely to ashes but for a few bits like the hands or feet. Even bones are burned away. The second type is a dinner party or some other event, where many witnesses all see one person suddenly go up into flames for no evident reason, and the flames are usually extinguished before the person can be killed. For each of these two types, there are a few very prominently repeated examples that you'll find on the Internet or in books. We'll give two examples of each.

Since this is my book, I'm going to invoke Author's Privilege and formally declare the two kinds. Spontaneous Human Combustion of the First Kind is when there are no witnesses to what happened:

- The most famous such case is that of 67-year-old Mary Reeser, whose remains were found by a friend in her St. Petersburg, FL home in 1951. Only her foot remained, still in its slipper, while the rest of her body had been reduced completely to ashes, along with the chair in which she'd been sitting. Her case is sometimes referred to as "the cinder woman".

- Another example you're likely to find in the books is that of 92-year-old John Bentley. A meter reader found Bentley's foreleg and his walker straddled atop a hole burned into his bathroom floor, and Bentley's ashes on the floor of the basement below.

Most sources cite something like 300 such cases of the First Kind, and they all follow this same basic pattern. A person, usually elderly, often overweight, frequently mobility challenged, is found burned almost completely to ashes, bones and all. Their surroundings show scorching but are usually not burned.

Now, in recent years, a pretty good theory has been publicized that adequately explains all (or most) reliably documented cases of the First Kind, and that's the wick effect, of which a candle is the most familiar example. The flame on a candle's wick is small, but its temperature is very hot; thus it has a powerful melting effect within its tiny sphere of influence. This melts the wax into liquid, which is drawn up the wick, where it vaporizes and burns. The wick itself does not burn due to the cooling effect of the vaporization; but once the wax is gone, the wick burns away as well.

The application of the wick effect to human corpses is not supposition, but proven fact. In 2001, the *Journal of Forensic Sciences* published an account of a test performed at the State of California's Bureau of Forensic Services in which a pig carcass was wrapped in a blanket and provided with a source of ignition. After a number of hours, the smoldering fire was extinguished and it was discovered that the part of the pig that had burned so far, bones and all, had been reduced to ash. The experiment was repeated on the BBC television program *QED*. The body burns very slowly, with only a tiny flame or even no visible flame at all; and like a candle, the heat is so localized that very little else in the vicinity is affected by it.

In 1991, a pair of hikers in Oregon came across the body of a murdered woman in which a wick effect fire was still taking place. The middle portion of her body had been completely burned away, including the pelvis and spine, while the slow-burning smoldering fire was still taking place in both legs and the upper torso. The killer was later captured and confessed to having lit the corpse on fire using lighter fluid. Like the other victims, the woman was overweight, with a high fat content that is believed to have provided ideal conditions for the wick effect to take place. Mary Reeser and John Bentley were both overweight, had been wearing flammable clothes (exceptionally flammable in Mary Reeser's case), and both were smoking at the time of death. The condition of both corpses and the rooms in which they were found was perfectly consistent with what we'd expect to find if the wick effect had occurred.

However, not everyone accepts the wick effect explanation. Author Larry Arnold is among its most vocal critics. His 1995 book *Ablaze! The Mysterious Fires of Spontaneous Human Combustion* asserts that the cases mentioned above, and many others, have no natural explanation. Arnold wrote two other books as well, *The Reiki Handbook* about energy healing, and a report on what he believed were the psychic causes of the Three Mile Island nuclear accident. His explanation for SHC follows the same type of fringe reasoning. Arnold proposes that a particle that he called a "pyrotron" strikes the victim's body and ignites it from within. His pyrotron is unknown to science, of unknown origin, is undetectable, and has no describable properties, except that it seems to have something to do with Kundalini yoga. Though the publication of his book has made Larry Arnold something of a go-to guy expert on SHC, his explanation is clearly unacceptable from any reasonable scientific perspective. Science does not allow simply making up a subatomic particle and calling that a mechanism for anything.

Other researchers have proposed various other explanations, including methane, which is one byproduct of bacterial action in the gut. And, as every college student with a cigarette lighter knows, it's flammable. One problem with the methane hypothesis is that cows produce even more gas than humans, and if it were true, we'd expect spontaneous cow combustion to be common. But we don't have any reports of this. One explanation for the discrepancy is that cows do not participate in the triggering activities. Cows don't wear a lot of flammable rayon acetate nightgowns, and their nighttime place of rest is a rarely an overstuffed chair beside a crackling fireplace. They spend much less time smoking than humans, as their hooves lack the manual dexterity needed to operate a cigarette lighter (explaining the lack of Saturday night barnyard hilarity).

Spontaneous Human Combustion of the Second Kind is when the event is witnessed and we have accounts of what took place. These accounts are quite different than those of the First Kind. Slow, smoldering fires are never the case; they are always a large sudden ignition with active flames. When the victims survive, the burns (which can be serious) are on the skin, never the deep, complete reduction to ashes seen in the First Kind.

- In London in 1982, Jeannie Saffin, a severely mentally handicapped elderly woman, was sitting at a table with family when her upper torso suddenly caught on fire. They extinguished the flames and paramedics took her to a burn unit, where she died eight days later of lung damage from inhaling the fire.

- In 1938, also in London, 22-year-old Phyllis Newcombe's dress suddenly caught on fire as she was going downstairs at a dance. Other revelers extinguished the flames but she, too, died at the hospital from her burns.

People catching on fire is not especially uncommon. It happens all the time. The only thing differentiating the cases classified as SHC is that no source of ignition was found; the fires are said to have been spontaneous. Other than that, there's nothing especially remarkable about them. The fires burned in a familiar manner, and the injuries are what would be expected. But these cases of the Second Kind are also rare; probably more rare than the First Kind. The reason is that these are unsolved, whereas the First Kind cases are generally solved, at least to the satisfaction of the investigators. These two cases of the Second Kind are famous only because there was no source of ignition found. No cigarettes, open flames, or sparks were found near either Jeannie Saffin or Phyllis Newcombe; but it's not scientifically permissible to conclude that their combustions were spontaneous. Maybe they were; but just because we didn't find the cause hardly means that there wasn't one.

Structure fires or brush fires sometimes go unsolved as well, but I think you'll have a hard time finding a fire inspector who will invent the term "spontaneous structure combustion" as if the lack of a determined cause means there wasn't one. Spontaneous Human Combustion of the Second Kind should not be allowed to exist as a category; instead we should call them what they are: Unsolved deaths by fire. Similarly, SHC of the First Kind has never been found to be spontaneous either. Those are simply the rare cases where a natural death in isolation has been followed by a slow combustion from some nearby source of ignition.

The wick effect is an interesting tidbit of science, albeit somewhat gruesome. I find that the logical pitfall of calling either type of SHC "spontaneous", and instead recognizing why they're not, is even more interesting.

REFERENCES & FURTHER READING

Carroll, R. "Spontaneous Human Combustion (SHC)." *The Skeptic's Dictionary.* Robert T. Carroll, 24 Feb. 1999. Web. 10 May. 2011. <http://skepdic.com/shc.html>

DeHaan, J., Nurbakhsh, S. "Sustained Combustion of an Animal Carcass and its Implications for the Consumption of Human Bodies in Fires." *Journal of Forensic Sciences.* 1 Sep. 2001, Volume 46, Issue 5: 1076-1081.

Editors. "New Light on Human Torch Mystery." *BBC News.* British Broadcasting Corporation, 31 Aug. 1998. Web. 12 May. 2011. <http://news.bbc.co.uk/2/hi/uk_news/158853.stm>

McCarthy, E. "Fringe takes on Spontaneous Human Combustion, Gets Burned." *Popular Mechanics.* Hearst Communications, Inc., 1 Oct. 2009. Web. 6 May. 2011. <http://www.popularmechanics.com/science/4316466>

Nickell, J. "Fiery Tales That Spontaneously Destruct." *Skeptical Inquirer.* 1 Mar. 1998, Volume 22, Number 2: 15-17.

Palmiere, C., Staub, C., La Harpe, R., Mangin, P. "Ignition of a Human Body by a Modest External Source: A Case Report." *Forensic Science International.* 1 Jul. 2009, Volume 188, Issue 1: e17-e19.

10. THE SECRET HISTORY OF CHINESE MEDICINE

Westerners' belief that Chinese have long relied on alternative medicine is due in part to being duped by book publishers.

Today we're going to take a look at how Chinese alternative medicine spread into the Western world. Promoters of alternative medicine claim that this ancient wisdom was (and is) in common use throughout China, and the Western world is becoming aware of its value. Skeptics of this position point out that alternative medicine was only used in Chinese rural areas where conventional treatments were not available, and it became popular because it was inexpensive, not because it was effective. The actual history brings some interesting perspective onto both of these points of view.

So let's go back and visit revolutionary China, around the middle of the 20th century. Mao Zedong's Great Leap Forward was in full swing, precedent to the Cultural Revolution. At the beginning of this period, most Chinese were one of the world's isolated populations, to whose doorsteps modern innovations had not yet arrived, much like many Africans, Indians, and Indonesians. They lived largely unaware of what was happening in science and technology, and their worldviews were dominated by local traditions. Medicine was rarely seen by any of these populations; when someone was ill, traditional treatments based on centuries of unscientific beliefs were what was known and applied.

Meanwhile, throughout most industrialized population centers in the world, including China's big cities, hospitals practiced the leading edge of medicine. Chinese oncologists prescribed chemotherapy for cancer just like in other parts of the world. Patients in great pain would be given opiates. As early as 1949, the Chinese Academy of Sciences was one of the world's leading research institutions in life sciences and medicine. One difference that you would have seen between Chinese hospitals and those in other nations was the use of acupuncture, which was and still is in relatively wide usage; however, with an important proviso. In China, acupuncture is only used for pain relief, never as a treatment, and always in conjunction with conventional painkiller medications whenever available. It's essentially an ornament alongside the same basic treatments used in other modern hospitals.

This application of the best medical science available was all well and good for those Chinese living in the cities, but it wasn't doing much for those a thousand kilometers out in the country who were scarcely even aware that the cities existed. In 1958, Mao Zedong launched a Communist party magazine in China called *Red Flag*. Its title on the front page was written in Mao's own calligraphy. *Red Flag* was the government's primary mouthpiece throughout China for Mao's reform programs, explaining the plans and laying out the philosophy. One of the national problems that *Red Flag* addressed was healthcare. By 1964, the urban population was still less than 2% that of the national population, with the overwhelming majority living in remote rural areas. Yet that 2% living in the cities received almost all of China's healthcare budget, and all of the benefit of innovations from modern medicine.

Mao's government had tried since 1949 to recruit and encourage doctors to move from the cities to the country, but this had been largely a failure. What healthcare there was had been mainly provided by traveling teams of doctors who would spend a few weeks in the outlying provinces, but would then return to their hospitals in the cities where they could receive a decent income. The problem grew more pronounced with the increasing spread of schistosomiasis, a parasitic disease that caused infection and organ damage, and is most notably characterized by swelling in the abdomen. Schistosomiasis came to be something of an iconic symbol for the lack of healthcare in China.

Mao planned a fix. In 1968, *Red Flag* published what was to be China's solution. Mao's experience had taught him that efforts to push healthcare out to the countryside were doomed, and so he did the opposite. Farmers were recruited from all over China, given free training, and sent back to their own villages to serve as medical professionals.

Within just a few years, some 150,000 doctors and 350,000 paramedics — half a million workers to serve over half a billion patients — were at work throughout the country. They became known as the barefoot doctors. Candidates were required to be high school graduates. Most received three to six months of training at the nearest hospital, and when they worked as doctors in their villages they accumulated work points just as they did for their normal farming work. (Work points were a system used in communist China to tabulate each family's productivity, for which they could receive grain beyond the basic allotment, or other goods and services.) Barefoot doctors worked no more than half-time as medics; they were still required to continue their agricultural work to prevent productivity from suffering. Many barefoot doctors went on to later attend medical school and became licensed doctors. In fact, Professor Chen Zhu, China's Minister of Health and a Vice President of the Chinese Academy of Sciences, actually began his career as a barefoot doctor in the countryside in Jiangxi Province before deciding to go to medical school.

By 1970, China boasted a great army of half a million paramedics. They were trained and they were stationed throughout the country where they were most needed. But there was
still a problem, a very large problem. There were almost no resources to equip the barefoot doctors with medical instruments,
supplies, or drugs. There was a medical treatment available for
schistosomiasis, but there was no money for the barefoot doctors to provide it. So they employed the one resource China
had always had plenty of: Manpower. Schistosomiasis is caused
by worms, and these worms are spread by infected snails
through a local water supply. Throughout China, barefoot doctors directed workers to clear ponds and streams, and eradicate
the snail population. It was quite successful; within fifteen
years, this simple technique reduced the incidence of schistosomiasis from ten million per year to just over two million, and
in some areas, it was nearly completely eliminated.

Another significant part of their training was in first aid, to
address injuries and other medical emergencies. Pre- and post-
natal care was also taught, as well as basic hygiene like washing
hands before eating. The barefoot doctors were taught to recognize the symptoms of conditions requiring medical treatment, and were trained to refer such patients to the nearest
hospital. But what about everything in between; illness not serious enough to warrant hospitalization, wellness care, and
simple treatable conditions? Barefoot doctors were enabled to
prescribe medications, but the problem was that medication
was hard to come by and often too expensive for peasants. Mao
knew he didn't have the funds to stock 150,000 new pharma-
cies throughout China. So, he provided an alternative.

The Revolutionary Health Committee of Hunan Province
published a textbook, *A Barefoot Doctor's Manual,* intended to
equip the unequipped barefoot doctor with everything he needed. The manual is amazingly comprehensive, giving instructions for how nearly any expected illness can and should be
treated. It covers basic anatomy, birth control, hygiene, and
diagnosis. Interestingly, it also anticipates the likely unavailabil-

ity of needed medical therapies. So as a supplement, the bulk of its content is about medicinal herbs: What they look like, how to collect and prepare them, and what conditions they are believed to treat.

When *A Barefoot Doctor's Manual* reached Western cultures, Chinese alternative medicine went from being a vague curiosity to being an all-out pop-culture fad.

But the book provided a somewhat flawed introduction. It did not provide an insight into what was happening in Chinese hospitals, nor even what most fully licensed medical doctors would have practiced. *A Barefoot Doctor's Manual* instead showed what the worst equipped Chinese medics would have to resort to under the worst circumstances. Westerners got a slanted perception of Chinese medicine from the book.

While it's true that *A Barefoot Doctor's Manual* advocates alternative therapies, it also recommends conventional medical treatment whenever available. For example, the manual describes the treatment for Japanese encephalitis. It recommends the use of acupuncture, mud packs, a compress using extracts of toad, and herbal teas. But it also gives the list of conventional drugs that should be given intravenously. Throughout the manual, almost every disease listed includes the conventional medical treatment that should be given whenever available, and the traditional treatment to be given otherwise.

However, I found out something quite interesting when I tried to verify this. There are a number of English translations of *A Barefoot Doctor's Manual* available. I was intrigued when I saw that the page counts varied widely. Some are as short as 372 pages, some as long as 960. I could only find one edition available in electronic form to allow searching and comparison, and it's the one that's been most widely published in the US. It's from The Running Press, and its full title is *A Barefoot Doctor's Manual: A Concise Edition of the Classic Work of Eastern Herbal Medicine.* Note that word "concise". Beginning in 1977, The Running Press used to publish a longer version of the book, which is now out of print. Its title was *A Barefoot Doctor's Manual: The American Translation of the Official Chinese Paramedical Manual.* By scanning through brief snippets of the original full-length text available on Google Books, I found that the complete conventional medical treatment for Japanese encephalitis was given. But searching through the "concise" edition, which I bought, all such mentions of conventional medicine have been deleted. *Some 600 pages of sound medical information were cut from the book for Western audiences.* What remains is essentially a list of Chinese traditional treatments, with nothing to inform the reader that the barefoot doctors ever relied on anything else. The edition was not simply made concise; it was carefully edited to give a skewed (and untrue) impression of what Chinese medicine is. The publishers changed it from a responsible paramedical manual into a promotion for alternative medicine under the guise of "ancient Chinese wisdom".

And so, when we look at the history as a whole, we find that alternative medicine did not represent what knowledgeable Chinese doctors would have prescribed, at least not since the dawn of science-based medicine; and so the argument that alternative medicine is valid because the Chinese use it, is false. And we also find that the skeptical claim that Mao promoted alternative medicine through the barefoot doctor plan because it was cheap is not quite true either. The plan was an honest attempt to provide the best available medical care, and it only fell back upon alternative therapies when nothing else was at hand...and unfortunately, this was the case...all too often.

REFERENCES & FURTHER READING

Daqing Z., Unschuld, P. "ChinaError! Bookmark not defined.'s Barefoot Doctor: Past, Present, and Future." *The Lancet.* 29 Nov. 2008, Volume 372, Issue 9653: 1865-1867.

Editors. *A Barefoot Doctor's Manual: The American Translation of the Official Chinese Paramedical Manual.* Philadelphia: The Running Press, 1977.

Editors. *A Barefoot Doctor's Manual: A Concise Edition of the Classic Work of Eastern Herbal Medicine.* Philadelphia: The Running Press, 2003.

MacFarquhar, R., Fairbank, J. *The Cambridge History of China, Volume 15, Part 2.* New York: Cambridge University Press, 1991. 651-652.

Valentine, V. "Health for the Masses: China's Barefoot Doctors." *NPR.* National Public Radio, 4 Nov. 2005. Web. 17 May. 2011. <http://www.npr.org/templates/story/story.php?storyId=4990242>

Yihong P. *Tempered in the Revolutionary Furnace: China's Youth in the Rustication Movement.* Lanham: Lexington Books, 2003. 47-48.

11. MILITARY DOLPHINS: JAMES BONDS OF THE SEA

Are dolphins actually trained to attack divers and place limpet mines on ships?

It was 1973 when one particular plot to assassinate the President of the United States was thwarted. Dolphins, secretly trained for rudimentary communication with humans, had been fitted with harnesses to hold a magnetic limpet mine, which was to be attached to the hull of the President's yacht. It was the first time the public became aware that dolphins were employed in this manner by the U.S. military.

This was, of course, not reality, but the plot of the fictitious movie *The Day of the Dolphin* starring George C. Scott as a virtuous marine scientist. It played upon the idea that dolphins are used by the military to conduct dangerous underwater attacks, such as planting limpet mines on ships and attacking divers. Most people have heard these stories anecdotally and we generally assume them to be true. And like most urban legends, the idea of attack dolphins is not entirely fictitious.

This topic is deserving of a solid disclaimer. Some of what follows has disturbing ethical implications, and many may find it to be offensive. In this chapter we're taking that for granted, and will discuss only the facts of what has happened. Be forewarned, and don't read any further if you don't want potentially unhappy memories.

Let's look at what's publicly known about military dolphins. The United States Navy Marine Mammal System currently employs bottlenose dolphins and California sea lions in three basic roles: guarding ships and ports against enemy divers, recovering lost hardware, and locating underwater mines. Throughout the program's history, marine mammals have deployed in 25 countries. Other mammals have been employed in the past, including pilot whales and belugas.

Dolphins and sea lions have advantages that are hard for navies to ignore. They swim far faster than divers, and are much easier and cheaper to deploy than remote underwater vehicles. They can dive hundreds of meters and return, with no concern about decompression, quicker than a human diver could even get suited up. Dolphins' underwater acuity is such that they can acoustically detect a 3-inch ball 200 meters away in complete darkness, and even discriminate between different kinds of metal. A dolphin's brain is famously larger than a human's, in part because so much of it is dedicated to processing sonar signals.

Dolphins are used to find underwater mines. Dolphins of the MK 4 MMS (Marine Mammal System) use their echolocation to find mines tethered to the sea floor. If they find one, they notify their handler by pressing a positive-response paddle. The handler then gives the dolphin a marker containing an electronic transmitter, which the dolphin carries in a nose cup. The dolphin attaches the marker to the mine or its tether, and Navy demolition teams will later clear it. Dolphins of the MK

7 MMS perform a similar task, but they specialize in mines that are sitting on the bottom or even buried in sediment. Dolphins of the MK 8 MMS are trained to identify safe corridors for the initial landing of troops on shore. Their most publicized exploit was in 2003. In the first month after arriving in Iraq, six dolphins investigated 237 objects and cleared 100 mines.

The MK 5 MMS employs California sea lions to find and recover lost objects. They are used routinely to recover things like practice mines used in training, which are equipped with electronic pingers to make them intentionally easy to recover for re-use. The sea lions can also recognize and find objects that have been lost, using their acute low-light vision. They carry a bite plate with which they can attach a recovery line to the object, and are even trained to tug on it to make sure it's fast.

The MK 6 MMS is where the James Bond part comes in. Underwater, there is no contest between a human diver and either a dolphin or a sea lion. The difference in speed, agility, and stealth makes it embarrassing to be a human being. While a diver is lumbering along, a dolphin or sea lion can be in and out, and do whatever it wants, before the diver is even aware that anything has happened. MK 6 uses both dolphins and sea lions as sentries, guarding ships and ports against hostile divers. Even divers using stealthy rebreathers or swimmer delivery equipment have little defense against the marine mammals' senses and physical performance. MK 6 was deployed in the Vietnam War and has been used extensively in the Middle East since 1986.

Sea lions in the SWIDS (Shallow Water Intruder Detection System) program take this one step further. When they spot an intruder and notify their handler using the positive-response paddle, they are given marking hardware. The sea lion returns to the intruder and attaches the marker to his leg. The nature of this marker is not made public, but a safe assumption is that it makes a barbed whale harpoon look like a safety pin. What the Navy describes as "other security assets" then home in on the marker and do whatever it is they do. A sea lion's whole body is adapted specifically to chase agile fish at high speed and make a lightning-fast catch, and my analysis is that any intruding diver in the vicinity of a SWIDS team has a big problem.

And that, friends, is the extent of what the US Navy says it's ever done with marine mammals. But today's official story might not be the whole story. I'll give you a moment to recover from that shock.

In June of 1977, *Penthouse* magazine published an article called "The Pentagon's Deadly Pets" that told the story of the late Dr. James Fitzgerald. According to author Steve Chapple, Fitzgerald, an expert engineer in the use of sonar, was at a Navy cocktail party in Annapolis, MD in 1964 when he first floated the idea of using dolphins for military purposes. Soon, he was working for the CIA in Key West, training not only dolphins but all sorts of beaked whales that can dive deeper, and other marine mammals. One primary mission was the attachment of magnetic satchels to the hulls of ships. This had at least two potential uses: First, the attachment of explosives that could be used to sink the ship. Fitzgerald's animals were never able to restrict their attaching to only desired vessels, however, and this was deemed too risky to friendly ships to proceed with. Second, listening devices or other monitoring systems could be attached to enemy vessels. *Penthouse's* source for the article was a former CIA employee, Michael Greenwood, who had worked with Fitzgerald in the dolphin program. Greenwood told Chapple that the CIA once took one of Fitzgerald's dolphins aboard a boat disguised as a rumrunner and released it through a special underwater porthole, where it successfully attached a listening device to the hull of a Soviet nuclear powered ship.

The *Penthouse* article was lent extra credibility when Fitzgerald immediately sued Penthouse International, Chapple, Greenwood, and others for libel, based on a charge in the article that he'd tried to sell the dolphin technology to six nations in Central and South America. Documents were produced that included Fitzgerald's sales brochures in Spanish; and the judge found that for the purposes of this subject matter, Fitzgerald was enough of a public figure that he was fair game for printed criticism. Fitzgerald had even discussed his work on the TV news show *60 Minutes.* Enough documentation was produced in the lawsuit to adequately evidence the events in the *Penthouse* article.

The closest Fitzgerald had ever come to physically attacking divers was to train dolphins to tear a diver's air hoses and rip off his mask, forcing him to surface. But this is only as far as we know from Fitzgerald himself and other official sources. When we expand our search radius, it turns out the US Navy might have done a little more. Soviet reports of *Spetsnaz* special operations teams state that two frogmen — saboteurs of a demolition group ironically called *Delfin* (Russian for dolphin) — were killed by dolphins in Vietnam in 1970, while attempting to place mines on the hull of an American cargo ship. Ever since that incident, *Delfin* members trained for combat against dolphins, and successfully killed dolphin aggressors off the coast of Nicaragua.

Following the publication of the *Penthouse* article and the subsequent lawsuit, Michael Greenwood (the former CIA employee) went on the TV show *Tomorrow* and told interviewer Tom Snyder that dolphins were in fact used to kill, being armed with either a knife or a charged gas injection device. Along with what he described as cruel treatment of the animals, this training for violence was, he said, what caused him to grow disillusioned with the dolphin program and resign. During the 1970s, American divers used a weapon called the Shark Dart, made by defense contractor Farallon USA. This was a long dagger-like needle with a CO_2 cartridge that would inject the victim (shark or otherwise) with compressed gas and literally blow them up. If this is what was used, it would tie together Greenwood's claims with the *Spetsnaz* reports.

Soviet dolphins were fitted with similar antipersonnel weapons. In 1998, London's *The Independent* interviewed well known dolphin expert Doug Cartlidge, who had been invited to the Ukraine to assess the former Soviet dolphin program. At the time, they were struggling and had only four dolphins left, but they showed Cartlidge around their museum and freely shared documents with him. Cartlidge reported that at the height of the Soviet program, their dolphins were also trained to find and mark enemy divers; except their marker contained a CO_2 needle that could be remotely triggered, killing the diver if and when the Soviets wanted to. They also employed dolphins as kamikaze torpedoes with remotely triggered explosives, and told Cartlidge that as many as 2,000 dolphins had been killed testing and developing this system. But the most bizarre Soviet marine mammal system was a dolphin paratrooper. A dolphin wore a harness attached to a parachute, and could be dropped from heights up to 3,000 meters. How the dolphin was meant to get out of the harness once in the water, or what its task would be, was not reported.

In 2002, the *Times of India* announced that the Indian Navy had completed preparations to deploy a dolphin-launched limpet mine, citing the dolphin's superior ability to swim long distances in rough water without being detected.

There are, of course, many more such stories to be found, but the most common ones you might run across on the Internet I dismissed for being poorly sourced. If there are classified aspects of the US Navy's MMS, they're classified, and we don't know about them. However a look at what's known to have happened in the past, coupled with decades of improved knowledge about marine mammals and better technology, suggests that today's aquatic animals in the service of Uncle Sam are probably more capable than ever.

References & Further Reading

Anonymous. "Delfin." *Spetsnaz Training.* KMS, 1 Jan. 1992. Web. 26 May. 2011. <http://www.spetsnaztraining.com/view/dolphin>

Chapple, S. "The Pentagon's Deadly Pets." *Penthouse.* 1 Jun. 1977, Volume 8, Issue 10, Number 94: 48-54.

Davison, J. "Aren't they cute? Except when they're trying to blow you up..." *The Independent.* 9 Jun. 1998, Newspaper.

Editors. "Navy set to induct dolphins for dangerous sea missions." *The Times of India.* 16 Dec. 2002, Newspaper.

Ervin, Circuit Judge. "James W. Fitzgerald, Appellant, v. Penthouse International, Ltd. and Meredith Printing Corporation and Meredith Corporation and Bob Guccione and Steve Chapple, Appellees." *US Law.* Justia, 7 Jan. 1981. Web. 25 May. 2011.
<http://law.justia.com/cases/federal/appellate-courts/F2/639/1076/364411/>

Gasperini, W. "Uncle Sam's Dolphins." *Smithsonian.* 1 Sep. 2003, Volume 34, Number 6: 28-30.

Olds, R. *SSC San Diego Biennial Review.* San Diego: United States Navy, 2003. 131-135.

US Navy. "US Navy Marine Mammal Program." *Space and Naval Warfare Systems Command.* United States Navy, 5 Dec. 1998. Web. 21 May. 2011.
<http://www.spawar.navy.mil/sandiego/technology/mammals/>

12. NEAR DEATH EXPERIENCES

A comparison of the effects of hypoxia to the reports of a brush with the afterlife.

Today we're going to float around the operating room, look down at our own body lying there on the table, hear the heart monitor switch to a solid tone, and learn first-hand what some believe goes on during a near death experience. When a small percentage of people are near death or are temporarily dead, either from an accident or during emergency lifesaving treatment, they report eerie experiences that they interpret as having crossed the threshold into an afterlife. Some authors and researchers have catalogued these reports and concluded that the experiences must have been real, while some skeptical researchers have found that the experiences are the natural and expected result of low oxygen to the brain. It seems the perfect place to point our skeptical eye.

A favorite starting point when examining such tales is the application of Occam's Razor. This states that the explanation requiring the introduction of the fewest new assumptions about our world is probably the true one; in other words, the explanation that best fits our understanding of the way the world works. The supernatural explanation for near death experiences (NDEs) requires the existence of an afterlife, heaven or hell or whatever you prefer to call it. To science, which has never found any reason to suspect life might continue after the death of the body, such a place would be a major new assumption about our world. But to many people with certain religious beliefs, such a place is a given and the afterlife is real, and is thus *not* a new assumption. So to a lot of people, Occam's Razor does nothing to settle this particular question.

Probably everyone will agree to some extent that the brain is capable of generating surprising experiences, such as highly realistic dreams. We've all had faint or dizzy spells, and these can be pretty dramatic episodes even though, to an outsider, nothing notable happened physically. A bit later we'll talk about how all of the major events of an NDE are created in the brain in certain experiments. In summary, even those who believe that NDEs truly represent a brush with the afterlife probably agree that every experience that characterizes one can also be attributed to a natural cause. Why, then, is there a tendency to insist that they had to be an actual life after death experience?

In 1975, Dr. Raymond Moody published *Life After Life*, which became the seminal work promoting NDEs as evidence of an afterlife. Dr. Moody is a strong personal believer in not only the afterlife, but also reincarnation, claiming that he has personally lived nine previous lives. In his books he's cited 150 cases of people who, after resuscitation, reported extraordinary experiences.

Let's take a look at the reports. First, the basics. Although it's rare for any two stories to be substantially similar, there are common themes. One the most familiar is the life flashing before the eyes, a quick fast-forward replay of either the entire life or important events, even long-forgotten events, commonly called a life review. Perhaps the most popular report is a bright light, warm and inviting. Sometimes this is combined with a feeling of floating through a tunnel. Some NDEs include an out-of-body experience, usually floating in the air and seeing one's own body below, being tended to by medics, sometimes reporting seeing things happen that could not have been observed from the body's position. People with physical limitations, blind, deaf, or paralyzed, usually find that their bodies are whole during these experiences.

Some people report positive meetings with deceased loved ones or religious figures such as Jesus or Muhammad. Just as often, however, people report terrifying encounters with monsters, hated people, or the devil. So while many experiences are euphoric, many are very much the opposite.

So the question is, can we group all of these things together in such a way as to find an undeniable pattern? Is there enough consistency and predictability that we can conclude with good certainty that such a thing as an afterlife must exist, and here is the probable experience you'll have as you cross over? It's unlikely. When *Skeptoid* looked at The Hum, a worldwide acoustic phenomenon, we found enough variation to conclude that there are probable many different causes that likely have nothing to do with each other. NDEs are similarly complicated by many unrelated causes of characteristic experiences: drug effects, hypoxia, trauma, brain abnormalities, and simple dreaming, just to name a few. We'd expect people coming out of all these conditions to report things very similar to NDEs.

Let's take a look at out-of-body experiences. You can search the Internet and you'll easily find dozens (if not more) stories where someone floated off the operating table and made observations about the room, actions performed by surgical staff, and even things happening outside the room. I'm not even going to list them because there are so many, and I'll grant that many of them sound undeniable, that the only possible explanation is that the person's consciousness and perspective was indeed outside the body. Having read a lot of these, I make three observations:

1. I know a number of anesthesiologists. They are not impressed by these stories. It is common for patients to be aware during general anesthesia. They remember many details of the people, objects, and procedures in the room. We absolutely expect some number of supposedly unconscious patients to report things that happened that a layperson would assume were unknowable. In fact, *The Lancet* published research in 2001 that showed nearly 20% of patients retained memories of things that happened when they were clinically dead.

2. What's rarely or never written up in books is the fact that most such "recollections" get their details wrong, and were probably just imagined by the patient. When authors compile stories to promote the idea of NDEs, they tend to universally exclude these; in fact the majority were never recorded anywhere to begin with. If out-of-body experiences are truly part of passing over into the afterlife, then they usually represent an afterlife of some alternate universe where everything's wrong.

3. Some of the stories can't be explained by either of the above. They include specific details that the patient could not have known. Sadly, all of these are anecdotal. They're very interesting and I wish we had more of them, and that controls had been in place at the time. Since they weren't, the scientific method requires us to shrug and say "Neat, but not evidence, let's do it better next time."

As an example of the value of anecdotes in suggesting directions for research, Dr. Penny Sartori placed playing cards in obvious places on top of operating room cabinets at a hospital in Wales in 2001, while she was working as a nurse, as part of a supervised experiment. Although she's a believer in the afterlife, and documented fifteen cases of reported out-of-body experiences by patients during her research, not one person ever reported seeing the playing cards or even knowing they were there.

Life review, euphoria, bright lights, and meetings with sacred personages have all been correlated with high levels of carbon dioxide in the brain. Research published in the journal *Critical Care* in 2010 found that over one-fifth of heart attack patients who went into cardiac arrest and were resuscitated, all of whom would have had high CO_2, reported these phenomena. But these patients were all also nearly dead; so the NDE correlates equally well with being near death as it does with the physiological condition. To find out which is the best correlation, we'd have to see whether an NDE can happen when one condition is present and the other is not.

It turns out that extensive research has been done to characterize a person's experience with loss of blood to the brain when there is no risk of death, by that patron saint of human experimentation, the US military. For 15 years, Dr. James Whinnery put hundreds of healthy young fighter pilots into centrifuges to understand what a pilot might experience under extreme gravitational loads. He put them in until they blacked out. Once they reached a point where there was inadequate blood flow to the brain, they lost consciousness; and among the frequently reported experiences were the following: Bright light, floating through a tunnel, out of body experiences, vivid dreams of beautiful places, euphoria, rapid memories of past events, meeting with friends and family, and more. The list is an exact match with the events attributed by believers to a brush with the afterlife.

What about the reverse? Are there reliably documented reports of NDEs from people who were near death, but whose brains had normal oxygen supplies? If there are, I was not able to locate any. This leaves only one group of conditions that can be consistently correlated with what we call a near death experience, and it's not nearness to death. It's a set of brain conditions that includes hypoxia, hypercarbia, and anoxia.

Other researchers have also found ways to produce the symptoms of a NDE without nearness to death being a factor. In 1996, Dr. Karl Jansen published his successful results of inducing a NDE using the drug ketamine. In 2002, *Nature* published research in which experimenters gave direct electrical stimulation to the part of the brain called the angular gyrus in the parietal lobe. Subjects reported being able to see themselves lying there from a vantage point near the ceiling, and were able to communicate what they observed as it was happening. Some brain surgeries, most notably those for epilepsy, produce very high rates of NDE reports from patients whose lives were not in danger.

But believers in the afterlife are quick to point out that just because the reported experiences have natural explanations, it doesn't prove that the supernatural explanation is not also true in at least some of the cases. That's true, of course. We'd love to have such proof. Most of the symptoms of NDEs, like seeing a bright light and feeling euphoric, are too vague to serve as proof of the afterlife. But one isn't, and that's the out-of-body experience. What science would love to find is a win in a controlled test, consisting of the disembodied consciousness successfully completing a task under controlled conditions. If the claims of the most interesting such stories are true, this should not be a problem. It hasn't happened yet — nobody's yet seen Dr. Sartori's hidden cards, or beaten any other similar tests — but here's to hoping that they do. We all hope that death is not the end. Perhaps someday someone will prove Raymond Moody right, and we can all look forward to a catlike nine lives.

REFERENCES & FURTHER READING

Blackmore, S. "A Psychological Theory of the Out of Body Experience." *Journal of Parapsychology.* 1 Sep. 1984, Volume 48, Number 3: 201-218.

Blanke, O., Ortigue, S., Landis, T., Seeck, M. "Stimulating illusory own-body perceptions." *Nature.* 3 Oct. 2002, Volume 419: 269-270.

Braithwaite, J. "Towards a Cognitive Neuroscience of the Dying Brain." *The Skeptic.* 1 Jul. 2008, Volume 21, Number 2.

Jansen, K. "Using ketamine to induce the near death experience: Mechanism of action and therapeutic potential." *Yearbook for Ethnomedicine and the Study of Consciousness.* 1 Jan. 1995, Issue 4: 55-81.

Kruszelnicki, K. "Near-death myth alive and kicking." *ABC Science.* Australian Broadcasting Corporation, 8 Mar. 2007. Web. 3 Jun. 2011.
<http://www.abc.net.au/science/articles/2007/03/08/1866095.htm>

Moody, R. *Coming Back: A Psychiatrist Explores Past-Life Journeys.* New York: Bantam Books, 1991. 11-28.

Van Lommel, P., Van Wees, R., Meyers, V., Elfferich, I. "Near-death experience in survivors of cardiac arrest: A prospective study in the Netherlands." *The Lancet.* 15 Dec. 2001, Volume 358: 2039-2045.

Whinnery, J., Whinnery, A. "Acceleration-Induced Loss of Consciousness: A Review of 500 Episodes." *Archives of Neurology.* 1 Jul. 1990, Volume 47, Number 7: 764-776.

13. THE HAITIAN ZOMBIES

Are Haiti's legendary zombies real, or just an imaginative fiction?

Should you find yourself in Haiti, beware of the man they call the bokor. Vodou tradition warns that he may secretly sprinkle a powder onto you, causing you to fall ill and quickly die — apparently. Doctors can examine you and find no pulse, no respiration. You'll be buried and your friends and family will attend the funeral. Then, one night, the bokor will come and exhume your body, and administer a mysterious antidote. Your body will wake up, but your consciousness and personality will be gone. Your body will be a dumb slave, eating, breathing, working, following the bokor's commands, perhaps even sold into labor; until it grows old, the joints fail, and not even the bokor's magic can keep it going. The remainder of your life will be as a zombie.

To study the phenomenon of zombies, we have to answer a number of questions. We want to know if the Vodou tradition actually does include zombies, or if it's merely a dramatized caricature invented by storytellers to frighten Western audiences. We want to know if this has ever actually happened to anyone, the way it's described. Finally, if these answers are all yes, we want to find out if there is a pharmacologically plausible explanation for what's said to take place.

The first question is easily answered. Vodou tradition does include zombies, and belief in zombies is nearly universal among Haitians. 50% of the population self-identify as Vodou, and most of the rest call it Catholicism, though it's really a syncretism of the Afro-diasporic roots along with what the French missionaries taught. It's polytheistic and includes both white magic and black magic, and its ceremonies often involve drug-induced trances believed to be possession by gods. Bokors (Vodou sorcerers) openly practice zombification, and though few Haitians have ever met a zombie, even fewer doubt that it happens.

The second question, whether zombification has ever actually taken place, is the most important, and one that's all too often taken for granted in such mysteries. Before trying to explain a strange event, first establish that the strange event ever actually happened. If there's one thing I've learned from *Skeptoid*, it's that that answer is no more often than you'd think. Are there actually any reliably documented cases of people being raised from the grave and living out their lives as mindless slaves?

The most familiar case is that of the man who is the subject of Wade Davis' 1985 book *The Serpent and the Rainbow*, the Haitian man Clairvius Narcisse. In 1962, Narcisse was about 40 years old when he checked himself into the hospital complaining of pain and difficulty in breathing. Two days later he was dead, and his official death certificate is on file. His body lay in the morgue for 24 hours (in unsurvivable refrigeration) until his family collected his body and buried him in the family plot. 18 years later, Narcisse suddenly identified himself to his sister in the street and told a shocking tale. He'd been conscious but unable to move or breathe during his entire stay in the hospital and subsequent burial. After three days underground, his coffin was suddenly opened. He was beaten, gagged, forced to take a hallucinogenic drug, and dragged away to face two years of slave labor on a sugar plantation, as one of many similarly imprisoned zombies. He reported being in a dream state with no willpower the whole time. Finally another zombie killed a captor with a hoe, and the zombies all escaped. Narcisse wandered for sixteen years, afraid to return home, convinced that his brother had been behind the plot. Then, upon his brother's death, he visited his sister.

Narcisse has maintained his story, and even appeared in the 2008 documentary film *When the Dead Walk*. When Narcisse's story broke in 1982, Dr. Lamarque Douyon, director of Haiti's only modern psychiatric facility, wanted to know if the man was genuine. DNA fingerprinting was not yet available, so he questioned Narcisse, his family, family friends, and neighbors, and established to his satisfaction that today's Clairvius Narcisse was indeed the same person who was born and grew up with that name. It seemed to him that whatever zombification drug Narcisse had been given, if that was indeed the explanation, would be an important discovery. So he consulted with an associate in the United States, Dr. Nathan Kline, who sent the bright young Harvard graduate student Wade Davis to find it and investigate.

An ethnobotanist by trade, Davis was convinced that he'd found a satisfactory pharmacological explanation for what happened: the poison tetrodotoxin. He purchased samples of zombie powder from several bokors across Haiti, and though their preparations differed, he found three common ingredients: charred and crushed human bones, various plants with stinging spines, and pufferfish. Pufferfish is famous for being the source of the expensive and risky sushi *fugu*, which if not properly prepared, can administer a lethal dose of the neurotoxin tetrodotoxin. It comes from a bacteria that grows within pufferfish, newts, toads, and a number of other animals. Although they've built up resistance, for humans and predators it's deadly. Tetrodotoxin blocks a sodium channel in nerve cell membranes, preventing the nerves from being able to fire any muscles. In sufficient doses, which for humans is very small, it stops all muscle activity, including not only the voluntary muscles, but also the heart and muscles that control breathing. The victim is conscious but as there is no pulse or respiration, he loses consciousness and dies within minutes from the lack of blood flow to the brain. There is no antidote, and it can only be treated with artificial respiration if administered in time. However the effect is temporary, and a few *fugu* diners have recovered after their apparent death, having received just barely enough dosage to wear off before the brain was lethally damaged.

Davis asked about Narcisse's report of having been beaten and drugged when he was exhumed. The bokors said that zombies are force fed a paste made of sweet potatoes, cane syrup, and a plant called Datura. Along with nightshade and henbane, datura has long been used in many cultures as a hallucinogenic drug. It contains scopolamine, hyoscyamine, and atropine. But as a recreational drug, it's quite dangerous, and would need to be administered by someone experienced in its use, such as a bokor.

And so Davis published his book with this thesis: that tetrodotoxin produces a state indistinguishable from death, but that bokors have some way of resuscitating the victims, who have suffered sufficient brain damage from hypoxia that they live out their lives virtually as vegetables. But it received much scientific criticism, which focused on the zombie powder. Authors Takeshi Yasumoto and C.Y. Kao wrote in the British journal *Toxicon* that although pufferfish was indeed in Davis' powder, only insignificant amounts of tetrodotoxin were; and since tetrodotoxin breaks down quickly in water or in alkaline environments, it was implausible for it to be delivered as a poison in powdered form. However, the bokor preparations were all inconsistent with one another; some might have been stronger. In 1984, Dr. John Hartung at Downstate Medical Center in Brooklyn fed Davis' zombie powder to rats, rubbed it on their skin, and even injected it into their abdomens, and observed no effect.

From assessing the criticism and responses, my analysis is that a very few zombie powders are likely to deliver a dose of tetrodotoxin, and most are not. If bokors do indeed attempt to poison victims with their powders, the vast majority of such attacks will fail. Either an insufficient dose will be delivered, or the victim will die. Only once in a blue moon is it likely for a zombification to result in both a lucky knockout and a resuscitation, and the level of brain damage will be correspondingly variable. It's plausible, but unlikely; and if it happens at all, successes are probably extremely rare.

Skeptical criticism of Davis also pointed out that the entire case of Clairvius Narcisse is based on a flawed presumption. Even with all the documentation and evidence, and all the research done by Dr. Douyon, there is no reason to identify Narcisse as the man who died and was buried.

Haiti is not known for its wealth. In 1962, a hospital stay was beyond the means of many residents. The hospital where Narcisse is alleged to have died, the Albert Schweitzer Medical Center, charged $5 a day for local residents, and a higher rate for non-locals. But poor non-locals got sick too, and it was not unheard of for them to check in under the name of a local to qualify for the lower rate. We have no evidence to rule out the likelihood that some unknown man developed a fatal kidney failure (which is what the lab tests showed as the cause of death), checked in using Clairvius Narcisse's name, and subsequently died.

It turns out that there is a perfectly plausible reason that the real Narcisse might have been just fine with this. In his younger days, he had been something of the family black sheep. He had a number of illegitimate children whose mothers demanded support, and some other bad debts, and was not well respected. In fact, during Dr. Douyon's interviews, he learned that the family considered Narcisse's sins to be the reason that the bokors had punished him with zombification. It all made sense to them. And, for a man with Narcisse's skeletons in the closet, seeking a change of scenery is hardly unheard of. Perhaps in his later years he had a change of heart and wanted to reconnect with his family. Considering the convenient circumstances, his zombification was a perfect cover story.

Obviously this theory is unproven, but consider the alternative: He survived two days, conscious but with no heart or lung function and no oxygen to the brain, 24 hours of which were spent in hypothermic refrigeration in the morgue. He then spent three more days underground, and all of this with failed kidneys that lab tests showed were fatal; then emerged in perfect health and managed hard labor at the sugar plantation for years. Is all of that more plausible than a deadbeat dad simply skipping town?

We could go on all day and examine the many stories like that of Narcisse, but the rest lack his documentation and are anecdotal at best, despite their abundance. Vodou practitioners drug themselves regularly and perform zombie rituals, and some probably go for extended periods under the influence of datura. Almost certainly, in Haiti's history, some have been poisoned by tetrodotoxin in zombie powder and survived, though any with sufficient brain damage to be called "zombies" had questionable utility as laborers and certainly would never have recovered. The grain of truth in the zombie mythology is a real one, but its popular portrayal — even within Haiti — is an exaggeration based on tradition and superstition.

References & Further Reading

CIA. "Haiti." *The World Factbook*. United States Central Intelligence Agency, 12 Jun. 2007. Web. 9 Jun. 2011. <https://www.cia.gov/library/publications/the-world-factbook/geos/ha.html>

Davis, W. *The Serpent and the Rainbow*. New York: Warner Books, 1985.

Ducasse, R. "Connaissez-vous Narcisse?" *The Full Zombie*. Reynold Ducasse, 8 Aug. 2008. Web. 9 Jun. 2011. <http://thefullzombie.com/topics/narcisse>

Farson, D. *Vampires, Zombies, and Monster Men*. London: Danbury Press, 1975. 66-85.

Garlaschelli, L. "Zombie Fish Eaters?" *La Chimica e l'Industria.* 1 Sep. 2001, Number 7: 71.

Hahn, P. "Dead Man Walking: Wade Davis and the Secret of the Zombie Poison." *Biology Online.* Biology-Online.org, 4 Sep. 2007. Web. 10 Jun. 2011. <http://www.biology-online.org/articles/dead_man_walking.html>

Yasumoto, T., Kao, C. "Tetrodotoxin and the Haitian Zombie." *Toxicon.* 1 Jan. 1986, Volume 24, Issue 8: 747-749.

14. ANASTASIA

*Did the Grand Duchess Anastasia survive her family's
1918 execution, and go on to live in the United States?*

It was one of the twentieth century's great mysteries. The beautiful young Anastasia, last surviving member of Russia's Romanov ruling family, was rumored to have escaped her family's midnight assassination. She made it to the West and lived under the name of Anna Anderson. Yet some didn't believe she was who she said she was. Did Anastasia survive, or was her life cut short at the age of 17?

Grand Duchess Anastasia Nikolaevna

Anastasia was the Grand Duchess Anastasia Nikolaevna of Russia, the youngest daughter of Nicholas II, Emperor and Autocrat of All the Russias. Following a series of economic collapses and military defeats in Russia, culminating in World War I, the February Revolution in 1917 saw Nicholas abdicate his throne in favor of a provisional government. The Emperor and his entire family, along with a number of staff, were imprisoned in several locations. Finally, on April 30, the party were transferred to a two-story house in Yekaterinburg which became known as the "House of Special Purpose". Anastasia was 17 years old. On the night of July 16, 1918, a kitchen boy was called out of the house on an errand, and once he was clear, a squad of nine Bolshevik secret police, eight Russians and one Latvian, entered the home, bearing a signed order from the Supreme Soviet in Moscow, each carrying a loaded revolver. The household was awoken, and told there was unrest in the town and it was dangerous for them to remain upstairs. All seven members of the Emperor's family, along with the four remaining staff, were collected in a single room in the cellar. Chief of the Bolshevik secret police, Yakov Yurovsky, suddenly killed the Emperor with a single shot, and then the others opened fire until all the victims were dead or wounded. Yurovsky and some of the others then took bayoneted rifles and repeatedly speared all the bodies, including Anastasia and the other children, who had not yet died. Yurovsky reported that it was a horrible, bungled, confused mess of an operation; in fact, even in that small room, it was half an hour between the first and last gunshot. Yurovsky discovered that the daughters, whose bodies resisted both bullets and bayonets, were wearing bodices made almost entirely of precious stones, so dense that they acted as body armor. It was a singularly bloody and brutish execution.

The assassins stood over the eleven mangled bodies. They were hastily loaded into a truck and dumped in a watery three-meter deep pit at a local mine. The bodies were stripped to discourage identification. Their clothing, found to be rife with hidden precious stones, was burned in a bonfire. But when word spread, Yurovsky decided a more remote location would be best, so as to prevent the discovery of the bodies and their use as propaganda. The next night, the Bolsheviks returned to move the bodies. What followed became a black comedy of errors. In the effort to raise the bodies from the pit and transfer them to a more remote mining area a few kilometers away, trucks, carts, and horses were employed. Various vehicles broke down and the parties were separated. Yurovsky's men constantly focused on looting the bodies. Onlookers were driven away, no matter the time of night. Trucks got repeatedly stuck in the mud. Yurovsky himself spent an hour laying in the mud after his horse fell on him. Nobody knew how to burn a corpse, so help was sent for. In the end, most of the bodies were dumped in a shallow pit dug beside the road and covered with railroad ties, but two of them — the young boy Alexei and a young woman, identified by Yurovsky as the maid Anna Demidova — were buried nearby. All the bodies were burned using kerosene, and the faces disfigured using sulfuric acid. Yurovsky ordered his men to forget everything that had happened, and to never speak of it to anyone.

These events are known in such detail mainly from Yurovsky's own written report. Clearly, the bodies were under tight control of the secret police force for many hours, and were moved and removed; and it seems inconceivable that any could have escaped alive and unnoticed. However, stories have persisted that that's exactly what happened. It is not beyond imagination that the execution was botched even worse than Yurovsky admitted in his report, perhaps even to the point that one or more of the family could have escaped.

And so it did not really surprise the world when a young woman turned up in Berlin at a mental hospital after a failed suicide attempt in 1920, and claimed to be Anastasia. She used the name Anna Tschaikovsky, and told a harrowing tale of having been secreted to safety by one of the Bolsheviks, and a subsequent marriage to a young man named Tschaikovsky. They escaped to Romania, where her husband was shot dead in the street. For an entire decade, Anna Tschaikovsky was Paris Hilton, Lindsay Lohan, and Jessica Simpson all rolled into one. Doctors were split on whether they believed who she claimed to be. Members of the Romanov family, some of whom had met the real Anastasia and some who had not, were also split. For a time she was even financially supported by Nicholas II's uncle, Prince Valdemar of Denmark.

Anna came to the United States in 1928 where her fame preceded her. Now using the name Anna Anderson, she lived in luxury at the expense of Anastasia's relatives, enough of whom continued to believe in her story. But the psychological problems that originally put her on the map persisted, and a judge committed her to an asylum for a year. She returned to Germany in 1931, with her celebrity diminished, but with an increased number of lawsuits mainly filed by greedy individuals attempting to secure what they hoped would be a large Romanov fortune.

In 1968 she moved to the United States to accept the assistance of one of her diminishing supporters, whom she married. Although her husband was wealthy, they lived in squalor, with piles of garbage and scores of cats. Her health and psychiatric condition declined, and Anna Anderson died in 1984 while living with her husband in a car as a fugitive from court-ordered institutionalization.

But all the while that Anna Anderson was living the life of Anastasia, others were doing so too. In 1964, two old nuns died at a convent in Russia's Ural mountains. Their gravestones were marked Anastasia and Maria Nikolaevna, a secret kept by the priest who took them in back in 1919. Little else is known about their identities or how they could have managed to escape and make their way into the mountains.

In 1963, Eugenia Smith, born in Romania and living in Chicago, published a book in which she detailed her own adventures as Anastasia. In her account, she regained consciousness in the cellar after having been left for dead among her family members. A neighbor took her to her own home and secretly brought her back to health. Along with a sympathetic Bolshevik soldier, they snuck out of Russia on foot and sought sanctuary in Serbia, from where she eventually moved to the United States. For thirty years she raised increasing support, mostly from Americans, who believed her to be Anastasia. Five years after her book was published, *Anastasia: The Autobiography,* she died in New York City in 1968.

And yet, these are only a few of the many claimants. At least half a dozen women have claimed to be Anastasia, no fewer than ten men have claimed to be her brother Alexei (and heir to the throne), and even a raft of people have "come forward" as the sisters Olga, Tatiana, and Maria, or various descendants of the same.

The 1991 excavation of the main gravesite did little to settle the claims of the world's many Anastasias. It did contain only nine bodies, as Yurovsky had said. Anastasia's little brother, the 13-year-old Tsarevich Alexei, was not found, and only three of the four Nikolaevna sisters. DNA analysis confirmed the identities, but there was no way to definitively identify which three of the four they were. Anastasia had been the youngest and shortest of the four, so methods such as skull measurements, descent of wisdom teeth, and collarbone maturity level were used by various researchers. In the end, the Russians reburied one of the bodies as Anastasia, but much scientific dissent felt that Anastasia was probably the missing body. Thus, the door for impostors (or even a potential real Anastasia) remained wide open. Perhaps there was no second gravesite; perhaps Yurovsky fabricated that in his report to cover up his failure to kill all eleven people.

The identification came thanks, in large part, to Prince Philip, husband of the Commonwealth's Queen Elizabeth II. His grandmother was the sister of Nicholas II's wife Alexandra. Thus he shares mitochondrial DNA with Anastasia and her siblings, as it's inherited from the mother. His great-grandmother was their grandmother. Prince Philip gave a tissue sample which allowed British scientists to calculate a 98.5% certainty that the bodies were indeed those of the royal family. But that 1.5% uncertainty, coupled with the fact that two of the children's bodies had never been found, allowed the claims that Anastasia had survived to persist.

Then, in the summer of 2007, a team of amateur archaeologists followed the description of events in Yurovsky's report. Using metal detectors, they combed through the land near the already-discovered gravesite, and at one point they got some readings, about 70 meters away. Working carefully, they dug and sifted. Ultimately what they found was a graphic illustration of Yurovsky's morbid tale. A container of sulfuric acid. Bullets. Nails and strips of metal from a wooden box. And the burned remains of two people, identified by an anthropologist as those of a boy between the ages of 10 and 13, and a young woman between the ages of 18 and 23. We now had eleven bodies, but were they the eleven we expected? Tissue samples were obtained and placed under the microscope of science. It took some time, but when the DNA results came back they spoke firmly. The bodies were indeed those of the young Tsarevich Alexei, and the fourth of his sisters, probably Maria. All four sisters were now conclusively accounted for. Without any meaningful doubt, the entire Nikolaevna family (including Anastasia) and their four staff had died on that night in 1918, and spent 89 years underground.

Although the various impersonators perpetuated an intriguing mystery and made the twentieth century just that little bit more exciting, we shouldn't forget that their doing so cheated bereaved relatives, and honest believers, out of a lot of money. And they reminded us of a lesson to remember: you should always be skeptical.

REFERENCES & FURTHER READING

Aleskeyev, V. *The Last Act of a Tragedy: New Documents about the Execution of the Last Russian Emperor Nicholas II.* Yekaterinburg: Urals Branch of Russian Academy of Sciences, 1996.

Attewill, F. "Remains of tsar's heir may have been found." *Guardian.* 24 Aug. 2007, Newspaper.

Coble, M., Loreille, O., Wadhams, M., Edson, S., Maynard, K., Meyer, C., Niederstätter, H., Berger, C., Berger, B., Falsetti, A., Gill, P., Parson, W., Finelli, L. "Mystery Solved: The Identification of the Two Missing Romanov Children Using DNA Analysis." *PloS One.* Public Library of Science, 11 Mar. 2009. Web. 17 Jun. 2011. <http://www.plosone.org/article/info%3Adoi%2F10.1371%2Fjournal.pone.0004838>

Gill, P., Kimpton, C., Aliston-Greiner, R., Sullivan, K., Stoneking, M., Melton, T., Nott, J., Barritt, S., Roby, R., Holland, M. "Establishing the identity of Anna Anderson Manahan." *Nature Genetics.* 1 Jan. 1995, Volume 9, Number 1: 9-10.

Ivanov, P., Wadhams, M., Roby, R., Holland, M., Weedn, V., Parsons, T. "Mitochondrial DNA sequence heteroplasmy in the Grand Duke of Russia Georgij Romanov establishes the authenticity of the remains of Tsar Nicholas II." *Nature Genetics.* 1 Jan. 1996, Volume 12: 417-420.

Klier, J. *The Quest for Anastasia: Solving the Mystery of the Lost Romanovs.* Secaucus: Carol Publishing Group, 1997.

Massie, R. *The Romanovs: The Final Chapter.* New York: Random House, 1995. 146.

15. CONSPIRACY THEORISTS AREN'T CRAZY

We usually dismiss conspiracy theorists as crazy people; but that doesn't tell the whole story.

Today we're going to descend into the darkest depths of the human mind to learn what makes a conspiracy theorist tick; or, as some would put it, to learn why his tick seems just a bit off. Is there anything we can learn from the conspiratorial mind, and is there a method to its apparent madness?

The human brain evolved in such a way as to keep itself alive to the best of its ability. For the past few million years, our ancestors faced a relatively straightforward daily life. Their job was simply to stay alive. Like us, they had different personalities, different aptitudes, and different attitudes. This was borne out in many ways, but the classic example that's often used is that something would rustle in the tall grass. Some of our ancestors weren't too concerned, and figured it was merely the wind; but others were more cautious, suspected a panther, and jumped for the nearest tree. Over the eons, and hundreds of thousands of generations, the nonchalant ancestors were wrong (and got eaten) just often enough that eventually, more survivors were those who tended toward caution, and even paranoia. In evolution, it pays to err on the side of caution. The brains most likely to survive were those who saw a panther in every breath of wind, an angry god in every storm cloud, a malevolent purpose in every piece of random noise. We are alive today as a race, in part, because our brains piece random events together into a pattern that adds up to a threat that may or may not be real. As a result, we are afraid of the dark even though there's rarely a monster; thunder frightens us even though lightning is scarcely a credible threat; and we perceive the menace of malevolent conspiracies in the acts of others, despite the individual unlikelihood of any one given example.

Conspiratorial thinking is not a brain malfunction. It's our brain working properly, and doing exactly what it evolved to do.

So then, why aren't we all conspiracy theorists? Why don't we all see conspiracies all day long? It's because we also have an intellect, and enough experience with living in our world that we are usually able to correctly analyze the facts and fit them into the way we have learned things really work. It is, exactly as it sounds, a competition between two forces in our head. One is the native, instinctive impulse to see everything as a threat, and the other is our rational, conscious thought that takes that input and judges it.

Let's look at two examples that illustrate the ends of the spectrum. David Icke is a British conspiracy theorist best known for his claim that most world leaders are actually reptilian aliens wearing electronic disguises. When you pause a video, he points to the compression artifacting and asserts that it's a glitch in the electronic disguise. However, he's out in the world, he tours, he writes books, he has a family and is a member of his community. He's not locked in an asylum as we might expect from hearing his theory. The reason is that he's probably not mentally ill at all. His brain is doing exactly what it's supposed to. He sees a group of powerful men, and the instinctive part of his brain suggests a sinister purpose. Imagine yourself seeing the ministers of the G8, or some similar collection. A thought passes through all of our minds, something like this: "I bet they all know something I don't know. I'd love to hear what they were talking about. They're up to something." That's the same thing David Icke thinks. It's exactly what our brains evolved to do. Our brains all want to go there.

And then the intellect receives this warning, and analyzes it, based on its knowledge. We all have different knowledge built from different experiences. One who has had negative experiences with authority is likely to gauge this situation differently than one who has not. David Icke probably has some past experience that makes his intellect properly — if incorrectly — assign more credence to the threat than is necessary; overtly so, in his particular case. Most of the rest of us have rarely seen a news story where a secret collusion among world leaders was discovered, so our intellectual understanding of the world has good reason to reject this particular instinctive threat as being improbable. Thus we conclude that it's probably just a group of businesspeople doing what they have to do. We all fall somewhere along that spectrum, and all perspectives are the result of our brains properly doing their job.

Take another example, this time of the Big Pharma conspiracy. Our brains see a group of huge, profitable companies in the same industry, and the pattern is obvious. Instinct throws up its warning flag. They're up to something evil, they are a threat. That is the brain's normal and expected first response. Next comes the intellectual filter to evaluate the reality of the threat. Only this time, many more of us match the threat to things we've seen in the news. There *are* genuine conflicts of interest in the pharmaceutical industry. Sometimes the rules for getting drugs approved *have* been bent. Drugs *have* been taken off the market after initial findings of safety, many times. In fact, the average person on the street probably knows little about the pharmaceutical industry except for examples of such cases making it into the news. In this case, the knowledge that many people's intellect uses to drive its conspiracy filter is likely to give this one a score of "Plausible". Let's look at the results. Suspicion toward Big Pharma is fairly widespread, while suspicion of reptoid aliens controlling the world is quite limited. Most other conspiracies fall somewhere in between these two.

Whether it's David Icke seeing reptoids or your coworkers and neighbors shunning government approved drugs, it's the same thought process, and it's the brain doing its job properly. Like a classroom of students who all honestly studied hard yet still got varying scores on the test, our brains are going to be right sometimes and wrong sometimes. But they're all following the proper steps to get there; conspiratorial thinking is not necessarily, by itself, indicative of psychiatric illness.

To determine when a person is over the line and should be treated, psychiatrists often look closely at the context. Does the conspiratorial belief integrate harmlessly with this person's life, or does it dominate? Has it caused problems: loss of job, loss of spouse, loss of security, or caused sociopathic behavior? These are the types of things that differentiate a belief system from what we call an illness. A person who thinks Barack Obama's birth certificate is fake is not ill, but a person who obsesses over it to the point of driving away their friends and family could well be.

The diagnosis is often delusional disorder. It must be a primary disorder, which means that it's not a symptom of some other condition the patient may have, it has to be the primary psychopathology. There are six types:

1. Erotomanic, usually seen when the patient believes some famous celebrity is in love with them.

2. Grandiose, claiming to be famous or ingenious, sometimes claiming to be a real famous person who is actually an impostor.

3. Jealous, when they believe their partner is unfaithful when it's not true to the point of irrationality.

4. Persecutory, when they believe they're being spied on by the government or some evil force, often filing frequent police reports or making confrontations.

5. Somatic, when they believe they have some undiagnosed or unique medical condition.

6. Mixed, some combination of the above with none predominating.

Such people could benefit from treatment, usually a combination of drug therapy and psychotherapy. However, as we've discussed before on *Skeptoid*, getting them to agree to treatment at all is often the primary barrier. They believe their delusion is real. They will present their evidence to prove it until the cows come home. It's often impossible to get them to consider the possibility that the reality of what they perceive might be due, in any degree, to psychopathology.

But if you've had a conversation with a conspiracy theorist — and almost all of us have — you've met people who do *not* display symptoms of delusional disorder far more often than those who do. The ordinary conspiracy theorist is an intelligent, sane, and generally rational person. They are, in fact, unsettlingly less different from *you* than you may have thought.

Not all detection of purposeful agency sees something evil. For example, we now know that the sun, moon, planets, stars, and constellations are simply other bodies floating through space and doing their thing, much as our Earth does. But early human cultures, who lacked better knowledge, suspected them to be purposeful entities that existed only to influence humankind on this one particular rock. This brain function that kept our species safe from threats also formed the basis for pagan religions, the great polytheistic European cultures, and astrology. Note that astrology still thrives today. Astrology is psychologically similar to conspiratorial thinking. Both represent the healthy brain's perception of purposeful agency in ordinary phenomena, but one sees danger while the other sees comfort. All of our brains naturally take us there, and it is only our learned intellect that reins us back. We're all hard wired to experience a deep-rooted excitement at the thought of opening a fortune cookie, though most of us have learned to put little stock in the fortune. And if handed today's horoscope, few can deny that their brain will go straight to their own zodiac sign to see what it says. There is no need to be embarrassed about doing either of these. It's one of the things your brain is supposed to do.

So, embrace your inner conspiracy theorist. Learn to identify and understand your own conspiratorial thinking, and you'll be better prepared to comprehend the position of the next conspiracy theorist you talk with. When you dismiss someone as paranoid or crazy, remember: it might be *you* who's wrong.

REFERENCES & FURTHER READING

APA. *Diagnostic and Statistical Manual of Mental Disorders, Fourth Edition (DSM-IV-TR)*. Arlington: American Psychiatric Association, 2000. 323-329.

Coady, D. *Conspiracy Theories: The Philosophical Debate*. Hampshire: Ashgate Publishing, 2006.

Goertzel, T. "Belief in Conspiracy Theories." *Political Psychology*. 1 Dec. 1994, Volume 15, Number 4: 733-744.

Novella, S. "Hyperactive Agency Detection." *NeuroLogica Blog*. New England Skeptical Society, 22 Mar. 2010. Web. 26 Jun. 2011. <http://theness.com/neurologicablog/index.php/hyperactive-agency-detection/>

Shermer, M. *The Believing Brain: How We Construct Beliefs and Reinforce Them as Truths*. New York: Henry Holt and Company, 2011. 207-230.

Vedantam, S. "Born with the Desire to Know the Unknown." *The Washington Post*. 5 Jun. 2006, Newspaper: A05.

16. "CURING" GAYS

Do therapies intended to help gays become straight actually work?

Today we're going to point the skeptical eye at therapies designed to convert homosexuals into heterosexuals. Such programs make headlines pretty frequently for their controversial nature, with proponents positioning it as an important spiritual or psychological healing, and opponents charging it with being misguided, unscientific, insulting, ineffective, and completely unnecessary. Some of the answers depend on your point of view, but others can be answered for certain with sound science.

This is one of those issues that has all sorts of philosophical, moral, and religious facets. Those are all valid debates, but here on *Skeptoid*, we leave those to others. Today we're going to focus on just the science. Is homosexuality truly a psychological condition that requires treatment? Can any of the existing treatments actually change sexual orientation? Has there been any meaningful success? There is actual science here, though it may be buried deep under a layer of ideological propaganda and misinformation; and understanding the true science is the best first step in forming opinions or policy.

The formal name of all such programs are Sexual Orientation Change Efforts (SOCE), and they include all kinds of different methods. There are psychological approaches including behavior therapy, hypnosis, conversion therapy, and support groups. There are religious approaches involving prayer or atonement. In the past, change efforts have included electroshock therapy, lobotomy, and a variety of surgical procedures to remove various parts of the genitalia on both men and women, including complete castration. Only relatively recently has homosexuality become legal, and even today it remains illegal in some countries. Punishment has been severe, ranging from execution, to chemical castration as was given to the father of computer science Alan Turing, to years of imprisonment and hard labor as was suffered by the poet Oscar Wilde.

Today, some groups still consider homosexuality to be a psychological illness, and others consider it to be a religious sin. So there are two basic directions from which change effort proponents come. The first, that homosexuality is an actual mental illness, was almost universally accepted, in nearly all countries, until quite recently. It was principally codified as such in Richard von Krafft-Ebing's 1886 book *Psychopathia Sexualis,* which influenced the psychological profession of the time to consider homosexuality a disease. It could either be acquired or innate. Von Krafft-Ebing found only limited treatment options, of limited effectiveness. As a Roman Catholic, von Krafft-Ebing's work was largely inspired by religious beliefs which held that any sexual activity leading to pregnancy was generally normal and healthy (even including rape), while any activity not leading to pregnancy (including homosexual sex, masturbation, or any recreational non-penetrating activity) was generally deviant and perverted.

This was the prevailing view until Sigmund Freud presented a differing opinion in the first half of the 20th century. Freud believed that everyone was naturally bisexual, with some tendencies toward both heterosexuality and homosexuality, one or the other usually dominating. He did not consider either to be an illness by itself, and most of his work with patients focused not on trying to change this cause but on helping patients deal with the resulting effects: confusion, self esteem, rejection, inhibitions, or unhappiness.

It wasn't until 1973 that the American Psychological Association (APA) voted to remove homosexuality from the *DSM-II (Diagnostic and Statistical Manual of Mental Disorders).* The

DSM-III was given a section on ego-dystonic homosexuality, which is an inner conflict between one's attractions and one's own idealized self-image, and by 1986 even this was removed completely. Patients who have particular stress about their sexual orientation are now treated the same as any other patients who are under stress about something. This is now the standard in virtually all Western countries.

According to nearly all surveys, the vast majority of people who voluntarily seek sexual orientation change are white male Christians in the United States. The principal face of Christian change efforts has been Exodus International (since closed down), founded in the 1970s, with nearly 300 Protestant and evangelical ministries in some eighteen countries. When it began, its mission was to convert homosexuals into heterosexuals. Over the years they realized that this was a virtual impossibility (more on that in a moment), and so today they concentrate mainly on changing behavior; in other words, getting gays to simply *act* straight. They advocate celibacy for homosexuals, and heterosexual sex only for bisexuals. The Exodus mission statement is:

> *Mobilizing the body of Christ to minister grace and truth to a world impacted by homosexuality.*

Exodus' primary method for changing behavior is something they call reparative therapy. According to their web site, they believe that homosexuality is the result of the individual's unconscious attempts to restore a damaged relationship with his same-sex parent (97% of Exodus patients are men). Reparative therapy consists of forcing oneself to participate in manly activities, such as playing sports and hanging out with heterosexual men; flirting with and dating women in an assertive way, and avoiding them otherwise; and avoiding activities that Exodus considers to appeal to homosexuals, such as going to the opera or museums. In addition, joining a men's group at a Christian church is a fundamental of reparative therapy. One of

Exodus' closest partners is the American Christian evangelical nonprofit Focus on the Family.

At least one prominent group promotes itself as a science-based professional organization of psychiatrists and psychologists who disagree with the APA's finding that homosexuality is not a mental disorder. NARTH, the National Association for Research & Therapy of Homosexuality, charges the APA with politicking and creating a hostile environment that discourages research into homosexuality as a disease. NARTH acts as a referral service for more than 1,000 member counselors, mostly church groups. NARTH does not require that its members be licensed therapists or doctors. NARTH also works closely with Focus on the Family. From these facts, and from a keyword search of their web site and published biographies of their directors, it's hard to argue that NARTH is not a religiously motivated group, despite their omission of any overtly Christian objectives in their mission statement.

The overwhelming majority of psychological studies show that change efforts have never worked; however, both Exodus and NARTH link to a few cherrypicked studies in an effort to show that change efforts *are* successful. Chief among these is one by Dr. Robert Spitzer in 2001. It's a perfect example of how poorly conducted these studies are, and that it's one of the Exodus and NARTH favorites shows what a sparse field they have to work with. The main criticism is with Spitzer's methodology, a telephone survey of only 202 subjects, provided by Exodus. In fact, *fully a fifth of the subjects* were directors of Exodus International ministries. 100% of them were religious, nearly all Christian. The selection criteria was that all had undergone reparative therapy after self identifying as being previously gay — in other words, the only subjects chosen for the survey were those who already shown the effect the experimenter hoped to prove! If that's not enough, more than three quarters of the subjects had spoken in public in favor of sexual orientation change efforts. Among the questions was to self-report their own sexual orientation on a scale of to 1 to 7, 1 being fully heterosexual, 7 being fully homosexual. But amazingly, of the 202 subjects, only *eight* stated that their Exodus therapy took them from 5 or more to 3 or less, the only range that could reasonably be called a successful conversion. Of those eight, seven were Exodus ministry directors, and the eighth refused a follow-up interview. Yet Spitzer concluded from all of this that reparative therapy was effective. Spitzer's methodology and conclusions were roundly criticized, including directly by the APA itself. When it was finally published in 2003 by the *Archives of Sexual Behavior,* one of the journal's sponsoring associations, the International Academy of Sex Research, resigned because of the decision to publish.

Interestingly, there has been a tendency for some leaders of the major "gay cure" organizations to engage in homosexuality themselves. Two of Exodus International's founders, Michael Bussee and Gary Cooper, left the organization in 1979 to divorce their wives, and became a couple in a 1982 commitment ceremony. In 2010, a director of NARTH, Dr. George Reker, resigned after having been caught touring Europe with a hired male escort. Incidents like these show that at least some of the leadership of the organizations have been men with personal internal conflicts between Christianity and their own sexuality. We often consider former drug addicts to be well conditioned to be drug counselors, and it's perhaps equally appropriate for these men who have had their own issues to counsel other gay Christians. These incidents shouldn't surprise anyone; in fact, they probably shouldn't even be considered incidents. Relapsing is an expected consequence of many forms of counseling. It doesn't necessarily mean that the counseling itself is fundamentally flawed.

Neither does the religious nature of "gay cures" say anything about their effectiveness, and it's not a relevant line of investigation. Whether gays *should* seek change efforts, or whether homosexuality is sinful, are not science questions. To find out whether the change efforts work, we review the scientific research. And fortunately, that's already been done, in about as comprehensive a way as possible.

In 2007, under pressure from a growing evangelical movement promoting change efforts, the American Psychological Association announced the formation of a "Task Force on Appropriate Therapeutic Responses to Sexual Orientation" charged with assessing how best to deal with patients who express a desire to change their sexual orientation, and to evaluate the effectiveness of existing interventions. Their report was published in August of 2009. The abstract summarizes their conclusions:

> ...*Efforts to change sexual orientation are unlikely to be successful and involve some risk of harm, contrary to the claims of SOCE practitioners and advocates. Even though the research and clinical literature demonstrate that same-sex sexual and romantic attractions, feelings, and behaviors are normal and positive variations of human sexuality, regardless of sexual orientation identity, the task force concluded that the population that undergoes SOCE tends to have strongly conservative religious views that lead them to seek to change their sexual orientation. Thus, the appropriate application of affirmative therapeutic interventions for those who seek SOCE involves therapist acceptance, support, and understanding of clients and the facilitation of clients' active coping, social support, and identity exploration and development, without imposing a specific sexual orientation identity outcome.*

As anyone who lives and breathes has learned, trying to change people is nearly always destined for failure. There is no upside in changing from one person with one set of problems into a different person with a whole new set of problems; it usually just compounds all those problems together. Sexual orientation change efforts are no different. Recognizing their failure is neither pro-gay nor anti-gay, and it's neither pro-religion nor anti-religion. They simply don't work, and from the psychological data, they are both ill-advised and unnecessary. Many people face genuine problems and seek genuine interventions. For such people, we do have genuine psychological solutions that are proven to work. Trying to change who you are is not one of them.

REFERENCES & FURTHER READING

APA. *Report of the Task Force on Appropriate Therapeutic Responses to Sexual Orientation.* Washington, DC: American Psychological Association, 2009.

Krafft-Ebing, R. *Psychopathia Sexualis.* Stuttgart: Ferdinand Enke, 1886.

Lasser, J., Gottlieb, M. "Treating patients distressed regarding their sexual orientation: Clinical and ethical alternatives." *Professional Psychology: Research and Practice.* 1 Jan. 2004, Volume 35: 194-200.

Robinson, B. "Changing Gays and Lesbians: Ex-Gay and Transformational Ministries." *Religious Tolerance.* Ontario Consultants on Religious Tolerance, 9 Apr. 2006. Web. 1 Jul. 2011. <http://www.religioustolerance.org/hom_evan.htm>

Sanchez, C. "Straight Like Me: Ex-Gay Movement Making Strides." *Southern Poverty Law Center Intelligence Report.* 1 Jan. 2007, Issue 128.

Schaeffer, K., Hyde, R., Kroencke, T., McCormick, B., Nottebaum, L. "Religiously Motivated Sexual Orientation Change." *Journal of Psychology and Christianity.* 1 Jan. 2000, Volume 19: 61-70.

Spitzer, R. "Can Some Gay Men and Lesbians Change Their Sexual Orientation?" *Archives of Sexual Behavior.* 1 Oct. 2003, Volume 32, Number 5: 403-417.

17. ZENO'S PARADOXES

Greek philosopher Zeno proved that movement was impossible with a few simple paradoxes.

Even if you think you haven't heard of them by name, you'll recognize them. The most familiar of Zeno's paradoxes states that I can't walk over to you because I first have to get halfway there, and once I do, I still have to cover half the remaining distance, and once I get there I have to cover half of that remaining distance, *ad infinitum*. There are an infinite number of halfway points, and so according to logic, I'll never be able to get there. But it's easy to prove this false by simply doing it, which we can all do. So we have a paradox, a contradiction, something that must be true but which, clearly, is not. Does there exist a solution that adequately addresses the con-

tradicting phenomena? Some say there is; some say there is not.

Zeno of Elea was a Greek philosopher, born about 490 BCE, and was a devotee of Parmenides, founder of the Eleatic school of thought in what is now southern Italy. Zeno survives as a character in Plato's dialog titled *Parmenides,* and from this we know what the Eleatic school was about and where Zeno was coming from with his paradoxes. Parmenides taught (in part) that the physical world as we perceive it is an illusion, and that the only thing that actually exists is a perpetual, unchanging whole that he called "One Being". What we perceive as movement is not physical movement at all, just different interpretations or appearances of the One Being. Personally, I think they smoked a lot of weed at the Eleatic school; but Zeno was into this and came up with his paradoxes in order to support Parmenides' view of the world. Zeno's paradoxes were intended to prove that movement must be impossible, therefore Parmenides must be right.

He is believed to have developed a total of about nine such paradoxes, but they were never published. The most famous and interesting are his three paradoxes of motion:

First is the paradox of Achilles and the tortoise, who contrived to have a footrace. Achilles, knowing he was the swifter, gave the tortoise a hundred-meter head start. In the time that it took Achilles to travel the hundred meters, the tortoise moved ten, so that when Achilles got there he found the tortoise still had a lead. In the time it took Achilles to run those ten, the tortoise moved another meter. No matter how many times Achilles advanced to the tortoise's last position, the tortoise had crept forward a bit more by the time he got there. Even though Achilles would seem to be the faster runner, it was impossible for him to ever catch the tortoise.

Second and most famous is the so-called dichotomy paradox, in which we repeatedly rend in twain every distance to be traveled. For Homer to walk to the bus stop, he must get halfway there. Once arrived, he must travel half of the remaining distance, and so on and so on, with 1/8 the distance remaining, then 1/16, then 1/32, then 1/64; he will have an infinite supply of remaining distances to travel, and thus can never arrive at the stop.

The third is the paradox of the fletcher who finds that all of his arrows are unable to move at all. At any given instant in time, the arrow is motionless in flight. During that frozen moment, the arrow cannot move at all, since it has no time in which to do it. Time consists of an infinite succession of moments, in each of which the arrow is unable to move. Nowhere can we find a given instant in which the arrow has time to move, and so no matter how many such instants we have, the arrow can neither fly nor fall to the ground.

Zeno's paradoxes are often touted by some people as evidence that physics or science are wrong. If an ancient Greek philosopher can describe a simple situation, which our intuition tells us is obviously correct, it's easy for us to assign it more significance than we do the confusing jumble that is modern science. Why should we listen to Einstein, who gives us a lot of unfathomable equations, when Zeno's elegant fables prove that the physical world is not as science tells it should be? Given this line of reasoning, it's hardly surprising that Zeno has become something of a darling to some New Age supporters of a spiritual, not a physical, universe.

Famously, upon hearing the paradoxes, a fellow philosopher named Diogenes the Cynic simply stood up, walked around, and sat back down again. My kind of guy. His response may have been glib, but it elegantly refuted Zeno's claim. At least, it refuted the physical implications of the claim, it did not address the philosophical aspects; nor did it provide the mathematical solutions.

Zeno's paradoxes are an interesting intersection between mathematics and philosophy. Mathematically, it's trivial to calculate exactly when and where Achilles will overtake the tortoise, but the philosophical argument remains (apparently) intractable. Bertrand Russell described the paradoxes as "immeasurably subtle and profound". So philosophers have come up with some pretty interesting efforts to try and resolve this.

One such tactic concerns the Planck length, which is the smallest possible unit of length. Planck units are all based on universal physical constants, such as the speed of light and the gravitational constant. Philosophically it's reasonably accurate to describe the Planck length as a quantum of distance, the smallest possible unit. This means that there are a finite number of Planck lengths (albeit a staggeringly large number of them) along the racetrack of Achilles and the tortoise, and between Homer and the bus stop. There cannot be an infinite number of points, and so Homer will eventually be able to arrive. However, while this sounds like it might elegantly solve the paradox, it doesn't. It's not possible to force a quantum solution onto a geometric problem. A simple illustration of why this is so is to imagine a very small right triangle with its two equal sides each of one Planck length. The hypotenuse would have to be $\sqrt{2}$ Planck lengths, which is not possible. Planck doesn't apply here. Despite efforts to conclude otherwise, we *are* dealing with infinities here. Or... *are* we?

Intuitively, we understand 0.9 (0.9999999...) to be a value that forever approaches 1 but never quite gets there. This is fine as a concept and a thought experiment, but it is mathematically wrong. 0.9 does in fact equal 1; they are simply two different ways of writing the same value. It's easy to prove this to most people's satisfaction by dividing both values by 3. Both 1 ÷ 3 and 0.9 ÷ 3 equal 0.3, therefore both are equal to each other. Another way of looking at it is to consider the fraction 1/9, which is equal to 0.1. 2/9 is equal to 0.2, and so on, all the way up to 8/9 = 0.8 and 9/9 = 0.9, and we all know that 9/9 = 1. When we divide the number 1 into 9 equal slices, that top slice goes all the way up to exactly 1, a finite and reachable number.

If this spins your brain inside your skull, realize that you already accept many other interpretations of the same idea. Consider any other number whose decimal value is an infinite repeating series, say 3/7. It equals 0.428571 and we all accept that it equals 3/7, not "a number approaching 3/7 but that never quite gets there." It's two ways of writing the same thing.

It's the same concept when Homer takes his final step and places his foot down, completing his journey to the bus stop. He did not take a journey of infinite length. We can write an equation that describes how his final step consists of an infinitely reiterating series of smaller and smaller fractions, just as Zeno said:

$$\sum_{n=1}^{\infty} \left(\frac{1}{2}\right)^n$$

We in the brotherhood call this an absolutely converging series, and *contrary* to Zeno's understanding, it equals 1.

Another popularly proposed solution, particularly for the fletcher's paradox, involves time and speed. Zeno, charges his critics, only considered the distances and geometry involved; and since he left time out of his paradoxes completely, he also excluded speed, since speed is a function of distance and time. When a body is in motion, its position is always changing. Motion is fluid, it is not a ratcheted series of jumping from point to point. Consequently, at any given moment in time, a moving body has no single exact position. Zeno's conjecture, that the arrow is always frozen at some point, cannot be observed, reproduced, or computed, since that's not the way things move. Imagine taking a photograph of a moving object. There will always be some motion blur. No matter how fast is the shutter of your camera, even infinitesimally fast, there will always be some tiny amount of blur. There is no such thing as a moving arrow frozen in time.

Similarly, Zeno's computation that Achilles will never catch the tortoise also omits time. Zeno's premise assumes that each segment of the race, wherein Achilles advances to the tortoise's previous position, takes some amount of time; and since there is an infinite number of such segments, it will take Achilles an infinite amount of time. This is also wrong. As the physical length of each segment decreases exponentially, in a converging series, so does the time it takes Achilles to traverse it. Achilles' time to catch the tortoise is represented by a converging series that equals a finite number.

Achilles *will* catch the tortoise, because the very succession of segments proposed by Zeno add up to a finite distance that Achilles will cover in a finite amount of time.

Homer *will* reach the bus stop, because all of those infinitely compounding fractional segments are an absolutely converging series equal to a finite distance.

The fletcher's arrow is *always* in motion once it is shot, at no instant in time is it ever frozen with a fixed position from which it has no time to move.

So to summarize Zeno's paradoxes, they're basically word games that play upon an easily misunderstood mathematical concept. There is no paradox, because Zeno's math was wrong.

REFERENCES & FURTHER READING

Baez, J. "The Planck Length." *John Baez*. University of California, Riverside, 9 Feb. 2001. Web. 8 Jul. 2011.
<http://math.ucr.edu/home/baez/planck/node2.html>

Gardner, M. *Aha! Aha! Insight*. New York: Scientific American, 1978. 143-144.

Huggett, N. "Zeno's Paradoxes." *Stanford Encyclopedia of Philosophy*. Stanford University, 30 Apr. 2002. Web. 10 Jul. 2011.
<http://plato.stanford.edu/entries/paradox-zeno/>

Lynds, P. "Zeno's Paradoxes: A Timely Solution." *PhilSci Archive*. University of Pittsburgh, 15 Sep. 2003. Web. 9 Jul. 2011.
<http://philsci-archive.pitt.edu/1197/>

Plato. *Parmenides*. Dublin: Hodges, Figgis, & Co., 1882.

Russell, B. *Our Knowledge of the External World*. Chicago: The Open Court Publishing Co., 1914. 165-181.

Whitehead, A., Russell, B. *Principia Mathematica.* Cambridge: University Press, 1910.

18. THE ABOMINABLE SNOWMAN

How likely is it that the Yeti of the Himalayas is a real creature?

Of all the world's famous monsters and wild men, the Abominable Snowman of the Himalayas is among the most elusive. If it does exist in the world's highest elevations, it has the advantage of geographic isolation from human spotters, and is more likely to be able to survive in numbers without having been seen than would a Bigfoot or a Loch Ness Monster, said to live near populated areas. Many local Himalayan people are proven to have a matter-of-fact belief in its existence, as seen by the artifacts and temples of Nepalese Buddhists. Both the Russian and Chinese militaries have expended state resources to investigate it. The fossil record has proven the prehistoric existence of Gigantopithecus in Asia, an ape which appears to be an ominously close match for the reports. And of those adventurers who brave the icy altitudes over 6,000 meters, some have brought back high quality photography of footprints that look simian enough. So what is the truth about the Abominable Snowman? Is it proven one way or the other, or must science consider this to be an unanswered question mark?

There are two kinds of eyewitnesses to Yeti reports, just like to any other event: the honest and the dishonest. Some people lie about or exaggerate a car accident they saw; some may make up or exaggerate a Yeti sighting. They may even create hoax footprints or stage a photograph to fool someone else. Why would someone climb to six or eight thousand meters just to pull off such a dumb stunt? Probably no one would; but people who are there to climb a mountain like to have just as much fun with their friends as people down in the cities. Just as you might dig a fake giant footprint at the beach, so you might carve a perfect Yeti footprint and photograph it next to your ice axe. The most famous such photo, taken in 1951 and which you've probably seen, was taken by Eric Shipton, a notorious practical jokester, who's always avoided direct questions about the print. It's suspiciously perfect, with crisp edges and no indication it was made by a moving foot; and improbably shallow in the soft snow. Shipton and his companion claimed it was one of a long track of prints they came across which they also photographed, but which clearly bear no resemblance to the crisp footprint and which experienced mountaineers identify as a goat track. So hoaxes are a part of the game, but they're not the part that helps us learn whether or not the Yeti actually exists.

Honest reports are better, but often they leave the investigator little to go on; since many of them are honest misidentifications. A distant sighting could be a true Yeti, but it could also have been another climber, a rock, or a known animal. There are at least three species of bear that live in regions where the Yeti has been reported, all of which can stand on their hind legs. The Tibetan blue bear, the Gobi bear, and particularly the Himalayan brown bear all leave strange tracks in the snow and all will forage campsites. In fact, one Japanese researcher, Dr. Matako Nabuka, concluded that the Yeti never existed at all; that its local name meh-teh was just a mispronunciation of meti, one word for the Himalayan brown bear. His view is not widely accepted, but it does illustrate just one more vertex of the puzzle.

And so, honest reports of a sighting, even photographs, need to be testable if we're to use them to build a case. Sightings leave the investigator nothing to test. It's important to maintain an open mind, but the truly open mind is also open to the possibility that the witness was simply mistaken. What the open-minded investigator needs is evidence he can test, and the willingness to accept what the test reveals even if it conflicts with his preconceived notions.

That means physical evidence. There are, or have been, two decent pieces of physical evidence purporting to be remains of a Yeti: a scalp and a hand. Physical evidence is always best, since it can be directly tested; and when the evidence consists of actual animal remains, we generally have a pretty good chance at doing a genetic analysis. The scalp and the hand came from Pangboche monastery in Nepal. Legend says the founding monk lived in a cave, and friendly Yetis kept him supplied with food and water. When one of them died, the monk preserved and kept its scalp, and built the monastery to display the scalp and honor the Yetis' service. The hand was added to the collection later, but there is no record of when.

Pangboche

In the 1950s, these objects came to the public attention mainly through the efforts of some of the most famous (or infamous) names in Bigfoot hunting. A team led by cryptozoologist Peter Byrne, financed by oilman Tom Slick, visited the monastery a number of times. They were allowed to take samples of hair from the scalp, but were only allowed to look at and photograph the hand. At one opportunity, Peter Byrne secretly took two finger bones from the hand, replacing them with human finger bones. The hair and the bones were sent back to the United States. Although one might hope that having this physical evidence would have produced answers, it seems to have instead fallen victim to the personal rivalries and mistrust that seem to have always characterized the Bigfoot hunting community. A whole team of Slick's scientists examined the items, but they expressed apathy, disinterest, and frustration with the process. Only one primatologist, William Osman-Hill, reported that the finger bones may have been anything other than human, but it was not a convincing verdict. He felt they were human, but did note some of what he described as Neanderthal characteristics. Little could be told from two small finger bones, other than that there was nothing ape-like about them. Slick's teams also sent back a number of other hair samples and stool samples, but none of them ever passed scrutiny either.

Gigantopithecus blacki (model)

Many of Slick's team and other Yeti hunters have been driven by the conviction that the creature has an excellent potential match in the zoological kingdom, in Gigantopithecus. Gigantopithecus was a genus of three extinct species of great apes that lived throughout what is now China and Southeast Asia. By far the largest, and most recent, of these was Gigantopithecus blacki, which when standing erect, probably reached 3 meters (10 feet) and 540 kilograms (1200 pounds). These measurements are extrapolated from the only known fossils, which are a few thousand teeth and a handful of jawbones. These have been excavated from a few cave sites, but mainly found among Chinese traditional medicine shops where they've been collected and traded for centuries.

The teeth are clearly those of an herbivore, and its principal diet was probably bamboo. Microscopic examination of teeth has shown scratches consistent with the consumption of fruits and seeds as well. Gigantopithecus blacki's nearest living relative, the orangutan, is nearly completely herbivorous also; but it sometimes eats a small amount of insects, honey, and bird eggs. The last of the Gigantopithecus is believed to have gone extinct some 100,000 years ago, which is fairly recent so we would not expect much significant evolution to have taken place. Therefore, if we were to discover a surviving population, we'd expect to find them in the lush bamboo forests of lowland Asia which provide the food source they're adapted for, not thousands of meters above the tree line on Himalayan glaciers where the only herbivorous food source is tough lichen glazed on the surfaces of exposed rock.

What about its posture? We don't know for sure, because we've never recovered any pelvic bones; but we do have two decent clues. The first clue is its immense body weight. Gigantopithecus was far heavier than a gorilla or an orangutan, but both those species distribute their weight on all fours (though both are capable of standing erect when they want to). Gigantopithecus was even more likely to need to do the same thing. A minority of primatologists, however, have pointed to the shape of its jawbone as being broader at the rear, like a human's, to accommodate a vertical trachea. When we stand up, our trachea goes straight down; unlike an ape's which is more at an angle as it stands on all fours. But most agree that Gigantopithecus probably walked on all fours to support its immense weight, and likely stood intermittently to its full height to reach pieces of fruit or tender bamboo shoots.

The Yeti is said to always walk bipedally, and most reports of its tracks have it doing so for great distances across steep sloping snowfields. This discrepancy in locomotive capability, combined with the complete absence of a satisfactory food source, makes Gigantopithecus an unacceptable candidate for the Yeti, not a good one, as some cryptozoologists assert. Yes, they are both large furry animals, but that's where the resemblances end. It's been famously said of comparisons of Gigantopithecus to the Abominable Snowman that it isn't abominable, it doesn't live in the snow, and it's not a man.

A few years after Tom Slick's cryptozoologists harried the beast, a better pedigreed team gave it a shot. Mountaineer Sir Edmund Hillary and zoologist and TV personality Marlin Perkins traveled to Nepal to examine the artifacts at the Pangboche monastery. Financed by the *World Book Encyclopedia,* they brought a team of scientists along with them to directly examine the scalp and hand. Unfortunately the hand was immediately determined to be human, and attention turned to the Yeti scalp. This time the monks permitted it to be taken back to the United States, where it was found to be merely the skin from the shoulder of a Himalayan sheep. Hillary summed up his experiences searching for the beast thus: "I am inclined to think that the realm of mythology is where the Yeti rightly belongs."

And so how should we answer our original question: must science consider this to be an unanswered question mark? The answer to that is easy. Science, by its very definition, must consider virtually everything to be an unanswered question mark, the Yeti included. Science never gives absolute answers; science gives us our best answer so far. Right now, our best answer so far is that the Yeti remains unproven. We have no good hypothesis that would explain its existence, and no evidence that is both testable and that has passed testing. My conclusion to the existence of the Yeti is: Probably not, but it sure would be cool.

References & Further Reading

Buckley, M. *Tibet.* Guilford: Pequot Press, 2006. 213.

Craighead, L. *Bears of the World.* Stillwater: Voyageur Press, 2000. 79, 94.

Hall, A. *Monsters and Mythic Beasts.* London: Aldus Books, 1975. 100-115.

Kennedy, K. *God-Apes and Fossil Men: Paleoanthropology of South Asia.* Ann Arbor: University of Michigan Press, 2000. 99-110.

Meldrum, J. *Sasquatch: Legend Meets Science.* New York: Forge, 2006. 39.

Regal, B. *Searching for Sasquatch: Crackpots, Eggheads, and Cryptozoology.* New York: Palgrave Macmillan, 2011. 42-52.

19. THE HESSDALEN LIGHTS

Scientists seek a high-tech explanation for these ghost lights in Norway. But is the true cause much simpler?

Let us travel back to 1981, when people in Norway's Hessdalen valley began seeing something strange in their night skies. Colored lights that hung silently sometimes appeared above the valley. Curious onlookers found they were best seen from a slope at the north end of the valley, looking south, and within a few years UFO seekers started converging on Hessdalen. And ever since, there has been one of the lengthiest and most technical of all UFO investigations.

Some call them UFOs, some call them ghost lights, and some call them Earthlights. None of the researchers have a very good idea what they are, but Hessdalen's residents don't seem too concerned. Theirs is a small, quiet valley, well off the beaten path, and only a few hundred people live in its scattered village. In 1983, Dr. Erling Strand, a UFO researcher, initiated Project Hessdalen as a sort of central framework for all the information gathered by UFOlogists and other researchers. Strand was convinced that some unknown branch of physics was responsible for the strange lights, and he encouraged as many scientists as he could to join his project.

In 1998, Project Hessdalen constructed on automated measurement station on the popular slope, with a clear view of the sky looking south down the valley. It was filled with an astonishing array of technical gear. It included a number of optical cameras, but also magnetometers, weather gear, and equipment for measuring low frequency electromagnetic radiation. Curiously, the station also includes a random number generator, as a part of the Global Consciousness Project, which seeks to prove that collective human consciousness predicts significant events by altering the output of random number generators. Presumably, this is to test whether the appearance of the Hessdalen lights is some sort of psychic event.

What the automated station found is that the lights usually appear in the sky between 9pm and 1am, and more often in the winter. Many photographs have been sent to their web server, and can be seen at hessdalen.org. When they appear, the lights each maintain a constant color temperature, and its spectrum remains consistent.

The researchers dug in to try and track down this elusive new branch of physics. One of Dr. Strand's associates, astrophysicist and radioastronomer Dr. Massimo Teodorani, wrote of the many theories they explored:

> *Mainly the following possible causes have been considered: ionosphere activity, solar activity, cosmic rays, magnetic monopoles, mini-black holes, Rydberg matter, heated nanoparticles, piezoelectricity, [and] quantum fluctuation of the vacuum state. For none of these causes, except for some aspects of piezoelectricity, was it possible to find a successful proof.*

Teodorani himself favored a rather colorful explanation. His theory is a bit lengthy, but I'll paraphrase it. He believes that water seeps into fissures in quartz in the valley, and when it freezes, it exerts pressure on the quartz, causing a piezoelectric effect that produces an electrical current. He then jumps to the assumed presence of a hovering ball of plasma, presumably ignited by the piezoelectric current, and the balance of his complex theory involves the characteristics of the plasma. It binds with water vapor, aerosols, and airborne mold spores. It keeps itself in equilibrium due to complex interactions, and keeps its shape due to a presumed "cool coat" of water and ions.

This is not a useful explanation; it is, at best, a pretty far-out hypothesis. It strings together a number of fringe suppositions, few of which are plausible or have ever been observed. For one thing, what the Earth's entire history has told us is that when water freezes inside rock fissures, it cracks the rock. Not once has it ever been observed to leave the rock intact and relieve the energy by producing electrical current that creates hovering plasma balls. For another thing, natural quartz does exhibit the piezoelectric effect, but there's very little quartz in Hessdalen. It's nearly all schist and sandstone. While schist does contain bits of quartz, the schist itself is easily fractured by ice and would relieve any pressure. Finally, the piezoelectric effect in some natural rock is real but barely detectable. To my knowledge, there's no example of such tiny microvoltages causing anything to ignite, either in the cold, wet natural conditions or in a controlled laboratory.

What Teodorani did was not to look for the source of the lights that were observed; it was, in contrast, an effort that started from the assumption that some unique geophysical process was producing balls of light that floated up out of the ground. He then came up with some hypotheses about how he thinks this might be accomplished. Starting with an explanation, and working backwards to try and match the observations, is the opposite of the way good science should be done.

Even if Teodorani's hypothesis turned out to be true in every detail, it would still not be an acceptable explanation for the Hessdalen observations. People live throughout the Hessdalen valley, and none of the observations have ever included lights coming up out of the ground, or even any lights overhead; indeed, nothing's even been reported to have been witnessed from inside the valley itself. To see the lights, you have to climb up on the hill at the north end of the valley, and look south, across the top of the valley's length. Teodorani's hypothesis presumes that his plasma balls are in the valley and floating upwards; in contrast to the observations that simply state the lights are visible to the south of the observation point. A better explanation for why nobody in the valley has ever seen balls hovering around them then rising overhead is that's not where the lights are, and that their sources are elsewhere.

One piece of advice that Teodorani and the other Hessdalen researchers appear to have failed to heed is that given by Marsha Adams of the International Earthlight Alliance in her 2006 paper titled *Air Navigation Artifacts near the Hessdalen Valley, Norway.* She states:

> *Earthlight researchers must be vigilant for artifact lights. In addition to vehicle headlights, house and ranch lights, stars, planets, Fata Morgana mirages, and natural aerial phenomenon, an important source of artifact lights are commercial and private aircraft operating near observation sites.*

Using simple commercially available navigation software, she identified a number of air traffic corridors, VOR navigational stations, and local airports. These included one corridor that I found particularly intriguing: a corridor 18° North from the Tolga VOR, proceeding directly up the Hessdalen valley, straight toward the Automated Measuring Station, from exactly the direction it's facing.

As previously discussed in other chapters in the *Skeptoid* book series dealing with so-called ghost lights — the Marfa Lights in Texas, the Min Min Light in Australia, and the Brown Mountain Lights in North Carolina — there are known physical phenomena that can account for such reports. A better way to investigate the Hessdalen phenomenon would be to at least consider the causes found for these other lights. The Min Min light was proven to be a sort of thermal lensing effect, where warm air collected in ground topography underlaid cooler night air, and reflected lights from over the horizon; an effect we call a superior mirage. However the Marfa and Brown Mountain lights turned out to be misidentifications of conventional lights in a direct line of sight, probably distorted by similar lensing effects. Such thermal irregularities were found in both places. In the case of Marfa, they were car headlights along two prominent highways. At Brown Mountain, it was the headlight of a regularly scheduled locomotive on the plain far away, as well as some other lights. The majority of the images captured by the Hessdalen automated measurement station look exactly like the high-powered landing lights of aircraft following the corridor north from the Tolga VOR.

Trondheim Airport (TRD) is due north of Oslo, Norway. This route passes almost directly over Hessdalen. When the Hessdalen lights first started receiving attention in 1981, charter passenger service between Oslo and Trondheim was in its infancy, having begun with only a few flights in 1976. By 1982, Trondheim had opened its third terminal; and by the time Scandinavian Airlines acquired the local carriers in 2002, the Oslo-Trondheim route was the single busiest in all of Norway. Trondheim is about 40 nautical miles north of Hessdalen, and at that range the use of aircraft landing lights is discretionary. Some pilots will have them on, some won't; and weather often affects their choice. Note the automated station's finding that the lights usually appear between 9pm and 1am, when it's dark and the air traffic is active; and more often in the winter, when more pilots like to use the landing lights during flight amid clouds. Whether the Hessdalen Lights are mysterious balls of unexplained plasma or not, extremely bright lights matching the photographs would have been visible in the sky on many nights. If you've ever stood near the approach to a major airport and looked downrange at night, you've seen the sky dotted with strange, stationary glowing orbs.

At the very same time Project Hessdalen was building its automated measurement station, and UFO researchers were furiously weaving together complex hypotheses about plasma and nanoparticles and electromagnetism and mini-black holes, the skies above were filled with the largest number of these strange, stationary glowing orbs that Scandinavian Airlines had ever put up there.

By no means am I suggesting that aircraft landing lights are the cause of all the Hessdalen sightings, but I have found no mention by any of the project's lead scientists that any serious effort was made to match the lights to aircraft flyovers in an attempt to falsify this particular hypothesis. Indeed, a few of the photographs from the automated station show lights down at ground level. Highway 576 winds along the floor of Hessdalen valley, so it would also be necessary to attempt to correlate ground level sightings with automobile traffic, which is exactly how the Marfa Lights case was solved in Texas. The effects seen at Marfa and in Australia of distant light sources appearing to be nearby, as liquidy and hovering orbs, can be quite impressive and persuasive; and it would be presumptuous of the Hessdalen investigators not to exhaust that possibility. Since we have an excellent and proven explanation for identical light phenomena in other parts of the world, that explanation is perhaps more likely than the list of exotic science-fiction avenues of investigation listed by Dr. Teodorani.

Investigate with an open mind. It's tough, because it means you have to be open to the possibility that you have not, in fact, discovered whole new branches of physics.

REFERENCES & FURTHER READING

Adams, M. *Air Navigation Artifacts near the Hessdalen Valley, Norway.* Redwood City: International Earthlight Alliance, 2006.

National Geospatial Intelligence Agency. "Tolga VOR." *World Aero Data.* WorldAeroData.com, 1 Oct. 2007. Web. 5 Aug. 2011. <http://worldaerodata.com/wad.cgi?nav=TOLGA>

Radin, D., Nelson, R. "Meaningful Correlations in Random Data." *Global Consciousness Project.* Global Consciousness Project, 1 Jun. 2009. Web. 29 Jan. 2010. <http://noosphere.princeton.edu/>

Seckbach, J. *Life as We Know It.* New York: Springer-Verlag, 2006. 494-496.

Stolyarov, Alexander, Klenzing , Jeff, Roddy, Patrick, Heelis, R. A. "An Experimental Analysis of the Marfa Lights." *Society of Physics Students.* American Institute of Physics, 10 Dec. 2005. Web. 1 Mar. 2007. <http://www.spsnational.org/wormhole/utd_sps_report.pdf>

Strand, E. "Project Hessdalen." *Project Hessdalen.* Østfold University College, 18 Aug. 2000. Web. 2 Aug. 2011. <http://www.hessdalen.org/>

Strand, E., Teodorani, M. *Experimental Methods for Studying the Hessdalen Phenomenon in the Light of the Proposed Theories: A Comparative Overview.* Halden: Høgskolen i Østfold, 1998.

20. THE ZIONIST CONSPIRACY

Are the rumors true that Jews are planning to take over the world's governments and banks?

Today we're going to point the skeptical eye at conspiracy theories that claim Jews are trying to take over the world. There is not just one version of this, there are many; and in their various forms, they've been around for centuries. There's hardly been a moment in the past 2,500 years when some group somewhere has not been fomenting mistrust and suspicion of Jews and their motives: The Jews want to take over your government, the Jews want to take control of your banks, the Jews want to abolish your church. The accuracy of these claims is one thing; the history behind them is another.

Although the word Zion means many things to many cultures, it's usually a place of peace and unity, and cross-cultural brotherhood. However it's most often associated with the Jewish people in particular. In that lexicon, the word Zion typically refers to the "promised land", the homeland promised by God to the Jews according to Judeo-Christian canon. Zion can also refer more specifically to the city of Jerusalem or the location of Solomon's Temple, and sometimes to the Biblical land of Israel.

Historically, a Zionist was any person who fought for the establishment of a Jewish nation in Zion. This was finally fulfilled over the course of many bloody months from 1947 to 1949, as various nations fought over the partitioning of Jerusalem and the surrounding region. The nation of Israel has held a tenuous foothold ever since, and it remains the political and spiritual homeland of all Jewish people all over the world. Since its establishment, the mission of Zionists has been to defend and strengthen Israel, and to oppose challenges to its sovereignty; in short, Zionism is Zionist nationalism.

Some critics of Zionism frequently broaden the application of the word Zionist to include any people anywhere who express support for Israel. Suffice it to say that anti-Semitism is not your everyday bigotry. Its roots run deep, it is cross cultural, and it's been institutionalized as an official national policy by some of the world's greatest superpowers. Nazi Germany is the only most obvious example of anti-Semitism as policy, but it's hardly the only one. 500 years before Christ, in the time of ancient Persia, Xerxes ordered all Jews in his kingdom to be killed. Various Roman emperors and Greek kings ordered the Jews to be exterminated. While the Christians prosecuted their Crusades against Muslims and Jews, the Muslims were forcing Christians and Jews to either convert or be killed. In the 1300s, Jews were widely burned at the stake throughout Europe for "causing" the plague. In the 1400s, the Spanish Inquisition burned some 30,000 Jews for refusing to leave their country. But this list could go on and on ad nauseum. Jews have always been blamed for something, and were always at the receiving end of the genocide. There are scant examples in history of Jews doing the same to anyone else.

And yet claims of Zionist Conspiracy have always persisted, lack of evidence notwithstanding. The most significant evidence of the existence of the Zionist conspiracy to rule the world appeared in St. Petersburg, Russia in 1903. *Znamya* was a small, short-lived newspaper published by an extreme nationalist faction called the Black Hundreds, for whom anti-Semitism was a core value. *Znamya* serialized 24 articles over 9 issues of the paper titled *The Protocols of the Sessions of the World Alliance of Freemasons and of the Sages of Zion.* They were, apparently, the recorded minutes of that group's meeting which took place sometime in the late 1800s. The headline was "A Program for World Conquest by the Jews: Minutes of a Meeting of the Elders of Zion". Its articles covered topics such as economic war, methods of conquest, acquisition of land, a transitional government, propaganda, destruction of religion, organizing disorder, and the control of stock markets. Russian ultra-nationalist Pavel Krushevan, the publisher of *Znamya* and openly anti-Semitic, refused to give his source for the articles, other than to say they were received by him in French and were translated.

The Protocols of Zion, as they were commonly called, were widely translated and reprinted. They appeared in numerous Russian publications for the next 14 years, and then arose in the west. Britain, the United States, and Germany began publishing them around 1920, and they've been available in print somewhere ever since. Today they are still published throughout the Middle East, in Venezuela, in Malaysia, and Indonesia; most often in Muslim countries.

The first Western publication is believed to have been in the *Times of London* on May 8, 1920, under the headline "The Jewish Peril". While some readers were shocked at what they believed the Jews were up to, others were more skeptical; and one of the *Times'* correspondents finally uncovered a shocker. A year after the publication, in August of 1921, the *Times* published a follow-up called "The End of Protocols" in which it was revealed that much of *The Protocols of Zion* was a nearly exact copy of an 1864 book by Maurice Joly called *The Dialogue in Hell Between Machiavelli and Montesquieu*. Joly, a conservative French lawyer, wrote it as a satirical account of how Napoleon III was planning to take over the world. Large sections of it were plagiarized word-for-word in the original French, literally replacing the word "French" with "the world" and replacing "Napoleon III" with "Jews". As Joly's book was (and still is) available for anyone to examine, this revelation established conclusively that *The Protocols of Zion* were a hoax.

But who was the hoaxster? Pavel Krushevan himself, perhaps? Many authors advanced theories over the ensuing 70 years, but it wasn't until 1992 and the collapse of communism that certain Soviet archives were unsealed and made available.

Russian researcher Mikhail Lépekhine spent five years going through the records, and the story he unearthed was worthy of a modern spy tale.

In the opening years of the 1900s, the Imperialist Russian government correctly saw impending revolution as a very real possibility. The primary job of the Okhrana, the Russian secret police, was to stave this off as best they could; and in a war of ideologies, propaganda is usually the best weapon. Anti-Semitism was endemic in the Imperial government (indeed, discrimination against Jews was a national policy), and blaming the empire's problems on the Jews was a familiar tactic of the regime. Pyotr Rachkovsky was the head of Okhrana based in Paris, where he could keep an eye on the many revolutionaries who had sought safety in France. Around 1900, while researching propaganda campaigns, he came across *The Dialogue in Hell Between Machiavelli and Montesquieu*. Documents unearthed by Lépekhine show that Rachkovsky's office hired the author and political activist Matvei Golovinski to write something up showing that the Jews were behind everything the revolutionaries opposed. Golovinski wrote *The Protocols of Zion* using the text provided by Rachkovsky as a guide, and it was then delivered back to Russia by their agent Yuliana Glinka. She delivered it into the eager hands of publisher Pavel Krushevan, and the rest is history. Mikhail Lépekhine's findings were written up in the French magazine *Le Figaro* in August of 1999, and again to much greater attention in *L'Express* a few months later.

In what seems like an almost pitiful last gasp of a dying regime, as the revolution progressed, the Imperialists began calling the Bolshevik revolutionaries themselves Jews. The term Judeo-Bolshevism was invented to associate Jews with communism, and later became one of Adolf Hitler's pet phrases. The ironic footnote to all of this is that Golovinski, author of the most infamous anti-Semitic document in modern history, changed allegiances after the 1917 revolution and became a Bolshevik himself.

With *The Protocols of Zion* having been so thoroughly proven to be a hoax as early as 1921, its continued publication and endorsement by such world leaders as Hugo Chavez and Mahmoud Ahmadinejad suggests that belief in the Zionist Conspiracy is driven by something other than true history or evidence. Many scholars have identified three basic types of anti-Semitism:

- Ancient anti-Semitism, which despised Jews because they were poor and considered inferior. Their culture had a late start and few friends, and they never got a strong foothold.

- Modern anti-Semitism, on the other hand, is largely based on suspicion of the Jews trying to control the world economy. Since the establishment of Israel, suddenly Jews have been considered to be in a position of strength from which they can wield international power.

- Christian anti-Semitism, which has its roots in Jewish deicide. This is the belief that all Jews, everywhere, are responsible for the execution of Jesus Christ. This underlying theme has been the primary driver of European anti-Semitism for two thousand years.

Why were the Jews always on the receiving end of holocausts, crusades, inquisitions, and imperial edicts? Why are there no stories of Jews riding forth and exterminating *their* enemies? The Christians and the Muslims fought each other, and both fought the Jews; we don't seem to have ever seen the Jews fighting back. The simple answer is that throughout most of history, there has never been a Jewish homeland. For 2,000 years, there have been no Jewish kings to send armies against the Christians and the Muslims. There were no Jewish dungeons in which confessions of apostasy could be extracted. The Jewish culture has been largely a civilization of refugees ever since Biblical times. They are history's greatest scapegoats and toughest survivors.

So far, not a single one of any of the myriad Zionist conspiracies has ever come true or been evidenced to exist. The only foundations supporting the current Zionist conspiracies are profound, historically-rooted anti-Semitism, and hoaxed or nonexistent evidence. Centuries of oppression have marginalized the Jewish community and they remain a tiny minority in the world; hardly the likely suspects to rise and conquer the world's economies and governments.

Some conspiracies are, undoubtedly, real. But of the many conspiracy theories that claim to predict future events, like a Zionist world government, not one has ever materialized. To those who promote or fear any given Zionist conspiracy, consider that all of its predecessors has faded away unfulfilled. I urge you instead to find your own Zion, a place of peace, unity, and cross-cultural brotherhood.

But if you insist on having evidence for the Zionist conspiracy, the official logo for the 2012 Olympic Games in London — a grouping of jagged polygons — was said to spell out the word Zion. Watch out.

REFERENCES & FURTHER READING

Conan, E. "Les secrets d'une manipulation antisémite." *L'Express.* LEXPRESS.fr, 18 Nov. 1999. Web. 10 Aug. 2011. <http://www.lexpress.fr/culture/livre/les-secrets-d-une-manipulation-antisemite_817525.html>

Hadassa, B. *The Lie That Wouldn't Die: One Hundred Years of The Protocols of the Elders of Zion.* Portland: Vallentine Mitchell, 2005.

Handwerk, B. "Anti-Semitic "Protocols of Zion" Endure, Despite Debunking." *National Geographic News.* National Geographic Society, 11 Sep. 2006. Web. 13 Aug. 2011. <http://news.nationalgeographic.com/news/2006/09/060911-zion.html>

Joly, M. *Dialogue aux enfers entre Machiavel et Montesquieu.* Bruxelles: Impr. de A. Mertens et fils, 1864.

Lendering, J. "Ancient Antisemitism." *Articles on Ancient History.* Livius, 25 Jul. 2001. Web. 9 Aug. 2011. <http://www.livius.org/am-ao/antisemitism/antisemitism02.html>

Philo Judaeus of Alexandria. *Against Flaccus.* Alexandria: (Ancient publication), c.40.

Pipes, D. *Conspiracy: How the Paranoid Style Flourishes and Where It Comes From.* New York: The Free Press, 1997. 85-87.

21. Are We Alone?

Are we alone in the galaxy, or might we have been visited by other civilizations?

Radio telescopes scan the skies, and computers crunch the results looking for the patterns that might indicate an artificial signal coming from deep space. Alien hunters stand watch out in the desert, looking for lights in the sky flying over military bases. Both are looking for answers to the same question: Is our little civilization on our little blue planet alone in the galaxy; or are there others, like us, who want to meet us as much as we want to meet them?

Are there technological alien civilizations out there?

Most astrobiologists think so. The physicist Enrico Fermi, upon comparing the apparent lack of any evidence of visitation to the inevitably huge number of civilizations out there, once famously blurted out "Where is everybody?" The most famous attempt to answer this question is the Drake equation, when Frank Drake strung together seven relevant variables in 1961. Multiply them all together — the fraction of stars that have planets, the fraction of planets that develop intelligent life, the fraction of those who choose to send signals into space, and so on — and you'll get the probable number of technological civilizations out there that we might hope to meet.

The obvious problem is that our estimates on most of these variables are all over the map. At Frank Drake's SETI Institute (the Search for Extra-Terrestrial Intelligence), 130 scientists in every discipline imaginable pursue research in more fields than you could shake a stick at, all of which helps to refine our estimates on each variable. The best guesses these days run from around 1 to 10. But nobody at SETI pretends that we have a good handle on this. Frank Drake himself said the equation was useful only for "organizing our ignorance" on the question.

DO THE ALIENS EXIST AT THE SAME TIME AS US?

A factor that many people fail to consider is time. Think of the galaxy as a Christmas tree with blinking lights throughout its space. Each brief light pulse is the life of a technological civilization. While there may be many lights turned on at any one given instant, the chance of two adjacent lights being on at the same time is much lower. Even if we could look into the night sky and see a big bright indicator light for every world that has a civilization, remember that that information is hundreds, thousands, even tens of thousands of years old. If we launched a spaceship to any one of them, even if it could travel at some meaningful fraction of light speed, the chances of that civilization still existing by the time we got there are small.

Even civilizations that survive their nuclear age and manage not to kill themselves are still vulnerable to Mother Nature. Terrestrial killers like supervolcanoes and pandemics, and cosmic killers like asteroids, novae and supernovae, can all destroy the hardiest populations. No civilization lives forever, and on a 14-billion year time scale, very few will happen to live side-by-side at the same time.

COULD ALIENS GET HERE?

The problems of interplanetary travel are well known. The distances, time, and energy requirements are all well beyond our current technology. These problems are equally difficult for the aliens. Even if we grant that their physiology may be well suited to multiple-century hibernation, the fact remains that interstellar space is a resource-starved desert and all the energy needed to decelerate must be brought with them.

We should also consider whether the visitation mission was one-way or round trip. Our Pioneer and Voyager probes are certainly only one-way. Building a vehicle intended to visit another star system and then return would be orders of magnitude more difficult. If it were intended to land, it would need to provide for re-entry for not just the lander itself, but also for an entire launch vehicle capable of taking off, breaking orbit, and returning. A far more plausible plan for a round trip vehicle would be orbital only, as this would greatly reduce the energy requirements for the return. But it also limits the science that can be done, and would not allow for direct contact.

Exotic science fiction solutions that avoid the problems of travel, like wormholes and space folds, have been studied and we do have a journeyman's understanding of them. Traversable wormholes — shortcuts from one point in space to another — have been theorized, but would require the use of exotic matter that has only been hypothesized. Folding or distorting space around you (called a warp drive in *Star Trek* terms) also has interesting real-life hypotheses, but the problems include absurdly immense energy requirements even to transport just a few atoms, and the self-defeating restriction that creating a warp bubble to travel 100 light years must always be preceded by preparations *taking* at least 100 years.

Of course, just because we haven't solved these problems doesn't mean no civilization can solve them. But while this special pleading is a philosophical possibility, it's not a practical possibility according to what we've learned so far.

This applies equally to the supposition that a civilization might have colonized its neighboring star systems, thus escaping cataclysm, and surviving and propagating indefinitely; but the problems of interstellar travel and the relative scarcity of suitable planets make colonization just as unlikely as visitation.

WOULD ALIENS COME IN PERSON, OR WOULD THEY SEND A ROBOT SHIP?

Sending a robot ship is much more practical. It can deliver all the messages, greetings, and information that the living aliens could; it has no need to carry life support systems; and its much smaller size makes the energy requirements far more achievable.

Even if the aliens plan to travel in person, they'd probably do as we've done with Mars, Venus, the Moon, and all the other heavenly bodies we've visited: send robotic probes first. No lives are risked, and a better understanding of the environment can be learned before sending living beings.

WHAT WOULD THE ALIENS DO HERE?

The question of what aliens would do if they got here is pretty hard to answer. There's no way we can know, but we can guess based on what we'd probably do if we visited someone else. When we sent out the Pioneer and Voyager probes, we put as much information about ourselves as we could onboard: what we look like, where we are, and a golden record with some recordings of sounds on Earth. Our best guess is that visiting aliens would want us to know about them, and help us to reply.

A fringe belief here on Earth is that aliens have visited, but merely stacked some rocks into pyramids, or drew a pattern into a cornfield, then left. It seems unlikely that if we were to go to all the massive development and cost of deploying a probe to an alien civilization, that this would be the plan we'd choose. We'd want to know about them, and we'd want them to know about us. The mission most likely to be successful would be to simply land as much information about ourselves as possible. Visiting aliens would probably do the same thing; simply land information about themselves. No return, minimal risk of failure.

We would probably not expect to have the energy available for our probe to fly around, move rocks, or evade their version of fighter jets. Maybe later in our technological development we might; but our first attempts at contact were simply to send a golden record.

Would we know whether they'd been here?

This question is largely answered by whether the aliens' visit was one-way or round trip.

Let's say we detected an alien civilization, and decided to send a space probe. By the time the probe got there, a huge amount of time would have passed; and it's entirely likely that during that time, the alien civilization would have advanced enough to make our visiting probe obsolete. So it might be a pretty good gamble to not bother intending the probe to make a round trip, but rather to allow the alien civilization to respond with whatever newer technologies they'd developed in the interim.

Considering the cost (energy cost or financial cost), we could probably land dozens of one-way probes for what it would cost to develop and launch a single round-trip probe. These costs would be the same for aliens as well.

Given the technological difficulties and greater energy requirements of a round trip probe, coupled with the probability that a one-way lander would be the better way to do science and invite a response, it's probable that we would know the aliens had visited us. Chances are we would have their probe on display in the Smithsonian, and would have followed the instructions on *their* golden record to reply.

We've never discovered an alien golden record, or a one-way alien probe, or any other evidence that we've been visited. So it appears that no aliens have yet visited us with the easiest, most probable option: a one-way unmanned robotic probe.

COULD ALIENS SKIRT THE PROBLEMS OF PHYSICAL TRAVEL BY VISITING IN SOME OTHER NON-PHYSICAL FORM?

Some like to suggest that advanced aliens might astrally project themselves, or find some exotic non-corporeal way to visit. This might make interesting fiction but it bears little resemblance to what we've learned about the way the universe works.

In reality, there is a way to non-corporeally accomplish the most probable mission, to deliver information about your civilization to a neighbor. Don't spend trillions of dollars and hundreds of years to land a box of stuff; beam the information there now, at light speed, for minimal cost, via radio. Anyone intelligent enough to listen for signals has already thought about translation. There are universal and mathematical constants that can be used as reference points: the Fibonacci series, the value phi, atomic mass units, Avogadro's number, and so on.

All of which brings us back to our original answer: the least likely scenario is that we'll be visited in the flesh, and there's no evidence that's happened. What's more likely is that we'll be visited by a robotic probe, and there's no evidence that's ever happened either. The most likely scenario is that we'll hear from our interstellar neighbors on the radio, and for as long as we've been listening, we haven't heard that yet either.

And so, are we alone? I don't think so. We know these problems are really hard to solve. If any pair of civilizations out there has solved them, we weren't included in the party. That doesn't mean we won't be tomorrow, or next year, or in a hundred years. Keep an eye on the sky if you must; but if we're going to meet our neighbors, chances are it will be the radio telescopes that find them first.

References & Further Reading

Gowdy, R. "SETI: Search for ExtraTerrestrial Intelligence." *Astronomy: A General Education Course.* Virginia Commonwealth University, 18 May 2008. Web. 10 Aug. 2011. <http://www.courses.vcu.edu/PHY-rhg/astron/html/mod/019/s5.html>

Lemarchand, G., Lomberg, J. "SETI and Aesthetics." *JonLomberg.com.* Jon Lomberg, 19 Jun. 2005. Web. 9 Aug. 2011. <http://www.jonlomberg.com/articles/seti_and_aesthetics.html>

Matson, J. "Alien Census: Can We Estimate How Much Life Is Out There?" *Scientific American.* 10 Feb. 2009, Volume 301, Number 2.

Plait, P. *Death from the Skies!* New York: Penguin Group, 2008. 7-32, 67-101.

Press, W., Teukolsky, S., Vetterling, W., Flannery, B. *Nature.* Cambridge: Cambridge University Press, 1986.

Shklovskii, I., Sagan, C. *Intelligent Life in the Universe.* San Francisco: Holden Day, 1966.

22 ALL ABOUT FRACKING

*Hydraulic fracturing of natural gas wells incites pow-
erful emotions. How much of the hype is justified?*

There are few technologies today quite so popularly dis-
liked today as fracking, short for hydraulic fracturing, the practice of pumping high-pressure water into natural gas reserves
deep underground to break up the rock and make the gas easier
to mine. Fracking has been harshly criticized all around the
world as dangerous, and has even been banned in a number of
countries. There are charges that fracking uses toxic chemicals
which contaminate ground water supplies; that it causes earthquakes; that it's killing endangered species; that tap water in
fracking areas contains so much methane that it can actually
burn; and that mysterious illnesses have resulted from the poisonous chemicals it pumps underground. Sound scary? It
should. But how much of it is true? Fracking is a perfect place
to turn our skeptical eye.

The 2010 movie *Gasland* brought these claims (and many
others) to the public attention. *Gasland* painted a horrifying
and emotionally charged picture of conspiracy, profiteering,
environmental ruin, and the reckless wholesale poisoning of
people and animals by the drilling companies. The energy industry was quick to respond to the apparent slander, even post-
ing a web page called "Debunking Gasland" (and others) that
not only denied virtually all of the movie's factual claims, but
also was heavy on ad-hominem attacks against its maker, an
activist whom they describe as an avant-garde stage director
with no expertise in either geology or drilling. Whom should
the average person on the street believe? Unfortunately, they

generally only hear from one of these sources or the other, and rarely or never get the unbiased, science-based facts.

Natural gas is found in the pores of shale or coal, and it escapes through natural fractures. Surface deposits are relatively easy to recover with simple drilling, no fracking required; but for the deepest and richest deposits, those from about 1.5 to 6 km underground, high pressure reduces the number of naturally occurring fractures and the rock's permeability is insufficient to extract much gas. These deep, non-porous formations are where fracking makes sense. Shale beds are often less than a hundred meters thick, so fracking boreholes are usually drilled horizontally to extend through as much of the bed lengthwise as possible, sometimes reaching through as much as a kilometer. The boreholes are sealed with pipe. High pressure water, sometimes as high as 10,000 psi, is then sent down this pipe. This system acts just like a hydraulic ram, where great force can be applied over a large area by introducing high pressure into the ram from a small entry point. This force splits the shale apart, creating numerous small fractures usually about 1mm wide. Then, to prop these fractures open to allow the gas to escape, a "proppant" is added to the water, which is basically sand. Getting this sand into the fractures is the whole point of fracking. Once done, extraction wells going straight down into the shale bed are far more productive, as the gas now has many free escape routes.

Although fracking has been in use since about 1950, it's only been applied in a large scale to natural gas mining since about 2000. About 90% of natural gas mines in the United States access rock that has been fracked. Fracking greatly increases the output of mines, producing domestic energy and profits and all the inflammatory economic and political implications thereof.

So what problems does fracking create? The most dramatic and popularly reported problem is that of burning tap water: hold a match next to a running tap and the methane contained in the water (methane is the main component of natural gas) will burst into flames. *Gasland* and many other sources have asserted that this is due to fracking. The burning water is a fact; whether it has anything to do with fracking is another matter altogether. Like much in science, the facts are more complicated.

The first thing to understand is that water wells are shallow. The deepest private residential wells go perhaps a couple hundred meters, though most are much shallower. Fracking takes place kilometers deeper underground; and in most places, the fracked shale beds are separated from the surface watersheds by multiple rock formations of different types. There's little or no transference of anything — gas or liquid — between fracked layers and surface layers; they're simply too far apart and separated by too much rock.

However, the burning water is an undisputed fact. So where is this methane coming from, if not from fracking? As it happens, it's natural, worldwide, for anyone who has a well in a natural gas area. Natural gas is not found only in the deep shale beds, it's in shallower layers as well; so we always expect some gas to make it into well water in particular regions. But the mining of natural gas also has a few consequences that can force methane into aquifers. First, the underground changes in pressure can prompt methane to migrate from areas of high pressure to areas of low pressure. Second, poorly sealed natural gas wells can (and do) leak methane into adjacent strata. These poorly sealed wells are human errors that it's the responsibility of the driller to repair. Third, old abandoned wells do the same thing, but often without anyone repairing them. None of these problems are related to fracking, per se.

When the Colorado Oil & Gas Conservation Commission investigated the burning water of the well owner most prominently featured in *Gasland,* whose tap water was gray and actually effervesced, they found that his methane was naturally occurring and had nothing to do with any natural gas drilling. His water well had been drilled directly into a shallow natural gas deposit. Nevertheless, *Gasland* portrayed this as a consequence of fracking, which is wrong at two levels.

Such well owners do have steps they can take to eliminate the problems, whatever the source of their methane. The simplest and most effective is to have the well properly vented. Methane weighs half as much as air, and venting it to the atmosphere is the standard practice and generally solves everything. Well venting would still need to be done even if fracking had never been invented.

It is an established fact that methane in tap water is found in greater concentrations in areas that have been fracked. In 2011, a much-publicized Duke University study found that, on average, levels were 17 times higher in private wells within 1,000 yards of a drilling site. But while an attention-grabbing headline implies a causal relationship, the only thing we know for sure is that this correlation is exactly what we expect to find. In areas where there is natural gas, (a) it's going to be found in wells, and (b) energy companies are going to come there to drill. The study noted that no data exists of methane levels in the water before the mines existed, and so no reason to suspect that mining or fracking had any impact on the levels. The researchers found that 13% of the wells had amounts of methane that exceeded "action levels", meaning that the wells should be vented to remove the methane.

What about the claims that fracking pumps hundreds of different poisonous chemicals into the ground? Well it's true, sort of, but not the way it's portrayed. The main chemical is simple water, which makes up about 98.5% of most fracking fluid. About 1% consists of one of many different types of proppant, basically sand. The type of proppant chosen for each job depends on the geology. The rest of the fluid, the remaining fraction of a percent, differs all the time. Mainly it's lubricant for the pumping equipment, borehole, and fractures. The goal is to get the sand distributed into the fractures to hold them open, and without the proper lubricants, surfactants, and suspension agents like guar gum, the sand collects in places and creates blockages. Depending on the type of rock, acids might be added to dissolve scale and get more water in. Then there are trace additives of things like corrosion inhibitors to prevent the pipes from corroding, and bactericides for killing bacteria that can clog or corrode. Complete lists of all the fracking fluid ingredients are widely available on the web, as required by law, and anyone concerned about them should take a look. A great place to start is to Google "fracking fluid disclosure".

If you're concerned about the fluids used in a specific well in your area, I recommend a web site called *Frac Focus,* which lets you search any well and find out exactly which kind of sand and other compounds were used there. *Frac Focus* is a partnership between the industry and the Groundwater Protection Council, an association of state and local regulatory agencies.

So when we're talking about corrosion inhibitors and benzene and guar gum and bactericides, any reasonable person who drinks water from that area should be concerned. So do you listen to pop culture, which tells us the fracking fluid is toxic and goes directly into your drinking water; or do you listen to geologists and regulators who tell us that never the twain shall meet? The hardest part of understanding fracking for the person on the street is knowing who to trust. I put this question to a friend who's a geologist for Pennsylvania's regulatory agency, right in the middle of some of the United States' most active fracking, and she acknowledged the problem. The movie *Gasland* is clearly an unacceptable source of information, and similarly, the drilling industry's own PR omits any frank acknowledgement of risks and ongoing investigations. They both have strong propaganda motives. The consensus for the best unbiased information seems to be the US Environmental Protection Agency. If you hate Halliburton, as many do for whatever reason, you'll love the EPA; they've even posted the subpoena they sent Halliburton for failing to provide required information about their drilling operations (instead, Halliburton famously had an exec drink some of its fracking fluid at an industry conference). If you want an unbiased understanding of fracking in general, start your education at EPA.gov/hydraulicfracturing.

At this time, the EPA is in the midst of a major investigation into the safety of groundwater supplies that may be affected by fracking. Unfortunately it moves at government speeds; the investigation is due to last through 2012 with a report due in 2014. In the meantime, the good news is that the EPA has yet to document any confirmed groundwater contamination from fracking operations. Even the Duke University study found no evidence at all of any fracking fluid in any of the wells they sampled. However, there have been a number of cases of contamination from accidental surface spills, similar to what we see from virtually every industry that transports and pumps liquids.

A number of nations have banned fracking until these investigations are concluded, but the EPA has not yet found any reason to do so in the United States. Like so many technologies, fracking has broad economic and political implications, and consequently incites fiery emotions on all sides. Your choice is whether to jump into that fire, or to study what we've learned by testing. Whether you think we need the energy production, or how much money Halliburton makes, are different questions; and should not be conflated with the science. Let the pundits answer those questions, and let the science determine the safety.

REFERENCES & FURTHER READING

Admin. "Debunking Gasland." *Energy in Depth.* Energy in Depth, 9 Jun. 2010. Web. 4 Sep. 2011.
<http://www.energyindepth.org/2010/06/debunking-gasland/>

Department of Environmental Protection. *Methane Gas and Your Water Well.* Philadelphia: Commonwealth of Pennsylvania, 2009.

DOE. *Modern Shale Gas Development in the United States: A Primer.* Washington, DC: U.S. Department of Energy, 2009.

Editors. "The Facts About Fracking." *The Wall Street Journal.* 25 Jun. 2011, Newspaper.

Fischetti, M. "The Drillers Are Coming: Debate over Hydraulic Fracturing Heats Up." *Scientific American.* 12 Jul. 2010, Volume 303, Number 1.

Fox, J. "Gasland: A Film by Josh Fox." *http://www.gaslandthemovie.com/.* International WOW Company, 17 Dec. 2009. Web. 3 Aug. 2011. <http://www.gaslandthemovie.com/>

Osborn, S., Vengosh, A., Warner, N., Jackson, R. "Methane contamination of drinking water accompanying gas-well drilling and hydraulic fracturing." *Proceedings of the National Academy of Sciences.* 9 May 2011, Volume 108, Number 20: 8172-8176.

Puko, T. "Fracking ruled out as contributor to East Coast quake." *Pittsburgh Tribune-Review.* 6 Sep. 2011, Newspaper.

Saba, T., Orzechowski, M. "Lack of data to support a relationship between methane contamination of drinking water wells and hydraulic fracturing." *Proceedings of the National Academy of Sciences.* 9 May 2011, Volume 108, Number 20: 8177.

Stelle, E. "Gasland Debunked." *Commonwealth Foundation.* Commonwealth Foundation for Public Policy Alternatives, Inc., 21 Jun. 2010. Web. 9 Sep. 2011. <http://www.commonwealthfoundation.org/policyblog/detail/gasland -debunked>

23. THE MONSTER OF GLAMIS

Did a living beast actually haunt Scotland's Glamis Castle for the better part of a century?

Today we have a good old-fashioned ghost story, and it even comes from a famous, dark old castle in Scotland. Glamis Castle (pronounced "Glahms") is the ancestral home of the Earls of Strathmore, and has been since 1372, when King Robert II of Scotland gave it to Sir John Lyon. His descendants eventually acquired the title Earl of Strathmore and Kinghorne, and a Lyon of that title has ruled the castle ever since. The Queen Mother Elizabeth Bowes-Lyon, who died in 2002, grew up in the castle as a young girl, as a daughter of the 14th Earl of Strathmore. Thus Glamis' history has been a highly public and visible one that ties our modern world all the way back with medieval times. And throughout all those centuries, it's held a secret. Legend says that the secret has been known only to the Earl himself, and passed on from elder

son to elder son. The secret concerns a hidden room and a living beast known only as the Monster of Glamis.

Our task today is to look at what's known about the Monster, and come to an informed best-bet conclusion to whether it really ever existed at all; or indeed, may even still haunt the grounds today...

The castle has fabled ghosts, of course. Any castle worth its salt has its share of those. There is the Tongueless Woman, alleged to be the spirit of a girl mutilated by castle guards for illicit relations with the Earl. There is the ghost of Jack the Runner, a black slave boy, said to have been killed by the Earl and his mounted party by their dogs in a variation of fox hunting. The grounds have a White Lady, and the chapel has a Grey Lady. And one the early Earls — variously described as either the second or the fourth — is said to have demanded a game of cards on the Sabbath, and when none of his guests would accommodate him, the Devil appeared at the castle gate and played cards with the Earl all night long. Some say he still plays in a secret, hidden room, condemned to gamble for eternity for his sin.

But it was during the reign of the 11th Earl of Strathmore that the Monster first came into the world. The wife of the Earl's son Thomas George, who would later become the 12th Earl, gave birth to a son, heir to the title on October 21, 1821. The boy was named Thomas Bowes-Lyon, and sadly, history records that the infant died that same day, and the title eventually passed to the second son, Claude, born in 1824.

But the legend says that young Thomas did not die, though perhaps it might have been better if he had. He was described as so horribly deformed that to look upon him was to invite madness. The poor infant was sent to a special hidden room and kept out of sight, as they expected him to die quickly. But he did not. The Monster lived and grew to manhood, fed by a single trusted servant through a grate in his door. He was said to be fat and round like an egg, with no neck, and only small, nearly useless arms and legs. The author James Wentworth Day first made this description public in his 1967 book *The Queen Mother's Family Story*. His unnamed source, whom some researchers believe to be the Queen Mother herself, gave the following description:

> *His chest an enormous barrel, hairy as a doormat, his head ran straight into his shoulders and his arms and legs were toylike.*

The tale goes on to say that on moonless nights, the same trusted servant would walk the Monster along the castle's battlements, and that section is still to this day called "The Mad Earl's Walk". What did lumber along the ramparts late at night, bellowing its mournful regrets, and scaring the flimsier ghosts back to their effigies?

Another story from Wentworth Day's book is that sometime in the 1870s, a stonemason engaged in remodeling encountered something that frightened him, and perhaps even discovered the secret room. The castle was equipped with a telegraph, and the Earl was summoned back from Edinburgh. He questioned the stonemason, and then paid him a large sum to move to Australia.

Legend has it that the secret of the Monster is passed from father to son, from Earl to Earl:

> *In 1904, the 13th Earl is said to have told an inquiring friend "If you could only know the nature of the terrible secret, you would go down on your knees and thank God that it were not yours."*

> *His wife badgered the estate's mid-nineteenth century factor, Andrew Ralston, to reveal the secret to her, and he said "It is fortunate that you do not know it and can never know it, for if you did know you would not be a happy woman."*

When his son, the 14th Earl, was asked about the secret by his daughter, the Lady Granville and the elder sister of the Queen Mother, he said "You cannot be told; for no woman can know the secret of Glamis Castle."

Years later, she reported "We were never allowed to talk about it when we were children. My father and grandfather refused absolutely to discuss it."

Now there's one thing that should raise a big red flag for you when you hear talk of a "family secret". Real secrets are secret, and you wouldn't be reading about them in a book if they were so. At the bottom of this page you'll find a list of references discussing this so-called "secret". So, clearly, if there was a secret, there were always at least a handful of people who knew it (trusted servants and factors at least), and it sounds like just about everyone else knew *about* it. At least, enough of them to furnish James Wentworth Day with enough material for his book.

But Wentworth Day's account is suspect on at least one level. According to the Queen Mother (he says), her own grandfather had insisted that the family secret not be given to him, and so it never was. So it's hardly likely that he could have then passed it down to his son, her father. How, then, could she have learned of the Monster? And why do we have accounts of both her father and grandfather issuing warnings about the secret?

It's a question we could chew on, but it may not be necessary to do so, when we look closer at our source. James Wentworth Day wrote over forty books on various subjects, but one of his favorite genres was — you guessed it — ghost stories. His bibliography includes such works as *Here Are Ghosts and Witches* (1954), *A Ghost Hunter's Game Book* (1958), and *In Search of Ghosts* (1969). He was not so much an historian as he was a weaver of illustrious tales. Nearly all the accounts you'll read of the Monster of Glamis reference Wentworth Day's 1967 book.

That's not to say that he invented the stories at all; I'm merely suggesting that he compiled and codified the history of the Monster, the hidden room, and the accompanying "family secret" into the version we have today. It turns out that for each of these story elements, Glamis Castle's earlier history provides a more concrete precedent, and one that is, conveniently, out of reach for researchers.

Glamis Castle has, since its earliest days, had a tale of a hidden room with a gory secret. In 1486, the 2nd Lord Glamis offered sanctuary to a fleeing band from a rival clan called the Ogilvies. But he betrayed them. They were led to a room, the door was locked and barricaded, and it was over a month before anyone ventured to look inside. Only one Ogilvie was still alive, having survived by eating the starved corpses of his companions. He was killed, and the room was permanently bricked up and sealed, to conceal the crime of having betrayed a promise of sanctuary. Though no known proof exists of this crime, it is consistent with the actual history of the day. And so Glamis Castle acquired a family secret, and a secret room that remains hidden. The purpose of passing the secret from father to son might well be to insure that the room is never opened.

In 1882, the New York Times published an account, many times retold and altered, of a party of guests who heard about the secret room and hung fabric from every interior window they could find. When they looked outside, a few windows had been missed, usually described as two adjacent windows in the central square tower. Many versions of this story exist, and in a castle as expansive as Glamis, secret or inaccessible rooms should surprise no one.

Wentworth Day even had a source for the stern warnings about revealing the secret, a 1925 autobiography by A.M.W. Stirling called *Life's Little Day: Some Tales and Other Reminiscences.*

All of this leaves us with one item to explain: the origin of the story of the Monster himself. No history exists for young

Thomas Bowes-Lyon, so the unfortunate infant probably did die of his birth defects. But hanging in the castle is an enigmatic family portrait from 138 years earlier. It depicts the 3rd Earl of Strathmore with his three sons. The Earl had five children, but this portrait includes only the male members of the family. The Earl himself is seated in the portrait. To his right stands his son and future Earl John, and to his left are two boys, one perhaps a young teen and the other a small boy. Two dogs stand with them, and watching over is an angel winging about in the sky. Glamis Castle identifies that portrait as having been painted in 1683 by the artist Jacob de Wet; in fact, we even have a diary entry noting the payment for "2 great Pieces for my Lord and his 3 sonns". We also have a contract from 1688 for de Wet to create a number of other artworks for the castle. So the portrait's provenance is well established.

Portrait by Jacob de Wet, 1863

But there is still a problem concerning the number of boys in the portrait. The Earl's two younger sons, Charles and Patrick, were not even born until 1692 and 1695, respectively; nine and twelve years after their portrait was painted. And Charles, the middle son, is recorded as having died in childbirth.

This has fueled rampant speculation that Charles did not actually die, but was horribly deformed, yet politely "cleaned up" by the artist. The painting is almost certainly the original source of the rumor of the Monster, but could it also be evidence that he was real?

Well, it certainly could be, but there's a much more rational explanation. It was common in past centuries, when a family picture was an expensive and time-consuming project, to depict babies and young children as older. We know for a fact that de Wet referenced "3 sonns" when only one had been born. At the time of the portrait's commission, the Earl and his wife may well have planned for more sons, and so chose to add two children to the picture. Both were eventually born, but one sadly died, and is now remembered only by the strapping teenage image of the young man he may have become.

Much of Glamis Castle is now open to the public and is easy to see, but much of its history has been lost to centuries of interclan squabbles, bureaucracy, remodeling, and the gradual merging of fact and fiction. Plenty of evidence exists of the circumstances in which the Monster of Glamis might have lived out his tormented years; but of that he ever actually had them at all, we have mere legend. The pitiable dear beast now resides with the White Lady only in the twilight land of legend. We may get more answers one day, when archaeology tells all Glamis Castle's secrets. Perhaps a few poor unfortunates can be buried on that day, and help us all rest a little easier.

References & Further Reading

Dash, M. "The Monster of Glamis." *CFI Blogs.* Charles Fort Institute, 9 Jun. 2009. Web. 16 Sep. 2011. <http://blogs.forteana.org/node/75>

Day, J. *The Queen Mother's Family Story.* London: Hale, 1967.

Douglas, R. *The Peerage of Scotland.* Edinburgh: R. Fleming, 1764.

Hope-Simpson, J. *Who Knows? Twelve Unsolved Mysteries.* Nashville: T. Nelson, 1974. 135-141.

Maple, E. *The Realm of Ghosts.* New York: A. S. Barnes, 1964.

Smyth, F. *Ghosts and Poltergeists.* London: Aldus Book Ltd., 1975. 76-80.

Stirling, A. *Life's Little Day: Some Tales and Other Reminiscences.* London: T. Butterworth, Ltd., 1925. 326.

24. Brainwashing and Deprogramming

Can people really be brainwashed, and then deprogrammed back to who they were before?

They reached their peak in the 1980s, a few elite specialists called deprogrammers. Their hotshot skill was as a sort of psychological exorcist, an expert at reversing the effects of brainwashing. Young people who were believed to have been forcibly brainwashed into joining cults, criminal organizations, or fringe movements could be deprogrammed in intense sessions, and thus restored to their former identities and value systems. It was a radical element of society that nobody talked about much, the sort of thing that someone might tell you about in hushed tones. The idea that it's possible to be brainwashed is a frightening one, and the notion of heroic deprogrammers repairing the damage is pretty remarkable. But the skeptical mind is forced to wonder whether brainwashing is really as effective as depicted, and whether dramatic deprogramming is truly the only answer.

Hollywood brainwashing usually takes the form of a virtuous person converted against their will into a mindless assassin. *The Manchurian Candidate, The Bourne Identity,* and even *The Naked Gun* have all followed this familiar formula. Deprogramming often made real life more thrilling than the movies, with midnight abductions, duct tape imprisonment in a basement, and harsh, sleep-deprived, antagonistic grillings.

There's a difference between brainwashing and simply convincing someone of something. Let's say I have my own religion or club or whatever it is, and I explain the benefits to you;

and after some discussion, you eventually decide that it sounds pretty good and decide to join. That's not brainwashing. Brainwashing requires a few specific elements. First, the beliefs have to be pretty radical. There's no clear delineation of what this means, but generally it's something that's significantly beyond anything you would have come to believe on your own without my influence. Second, my efforts have to be systematic. A lot of mere convincing is systematic; telemarketing pitches, for example. A lot of sales pitches follow a written script that includes branches and responses to typical objections. The difference with brainwashing is that the system is designed to change your beliefs, not to sell you something or get you to vote for a particular candidate. Finally, there must be some aspect of the pressure that is forcible. If you have the option to stand up and walk out any time you want, my pressure is not considered brainwashing. The definition requires me to apply some kind of pressure that eliminates your free will to leave my sphere of influence. Brainwashing is a forced, systematic, radical change of belief.

Take a group like the Hare Krishnas. They convince intelligent adults to shave their heads, wear robes, and forego worldly possessions. That's pretty radical. And, their recruitment methods are absolutely systematic. However, they generally don't force this onto anyone, so it's not brainwashing.

Scientology is well known for its highly structured, methodical, and relentless recruitment. But for ordinary members, the beliefs they advocate are not at all radical; little different from Oprah-esque self-help Kool-Aid. Their bizarre space opera beliefs are kept from new members. It's not until you're already in and approach the cult-within-a-cult, the Sea Org, that their recruitment qualifies as true brainwashing. The Sea Org is notorious for confining and isolating new members, imposing the uniforms and cutting off ties to family and friends. Radical, systematic, and forced.

Patty Hearst is one of the most famous cases of brainwashing. A wealthy heiress, she claimed to have been kidnapped by a group of young political radicals calling themselves the Symbionese Liberation Army, with whom she later committed a number of bank robberies and killings. She was convicted of bank robbery despite her defense that she had been a victim of the Stockholm Syndrome, where hostages develop sympathy for their captors and tend to support them. But on the theory that she was not responsible because her indoctrination into the group did indeed qualify as brainwashing, her sentence was commuted by President Jimmy Carter, and later she was pardoned by President Bill Clinton.

Brainwashing has also been practiced in wartime, most infamously by the Chinese upon American POWs during the Korean War. Their goal was to convert the prisoners into spokesmen for communism. This was done in three basic steps, all under great physical duress and exhaustion. First was to break down the prisoner's sense of self, eventually getting him to admit that he's not a killer, not here to hurt people, that it was wrong for him to come to Korea as a soldier. Finally his identity was in crisis. The second step was to offer him the possibility of salvation, along with some small gift, like his first drink of water in days; he need only reject his former beliefs. The third phase was to rebuild the new self, with growing rewards as he embraced each new aspect of his "good" new identity. This was brainwashing at its most pure: radical, systematic, and forced.

After the war, knowledge of this brainwashing process reached the pop culture consciousness. Over the next few decades, through hippie culture and New Age, cults began popping up more visibly than ever before, and a lot of young people experimented with them. This new idea of "brainwashing" was the first explanation that many parents reached for: "My baby would never join that strange cult; he must have been *brainwashed*."

And when there's a need, the market is quick to provide a solution. Deprogrammers appeared on the scene, presenting themselves as hired gun intervention superheroes, sometimes charging tens of thousands of dollars. They promised to reverse-engineer the brainwashing and restore the former identity. The pioneer was Ted Patrick, whose own son was (as he described it) "psychologically kidnapped" by a cult. He was a lifelong activist who, when faced with this new threat, boldly created the whole subculture of deprogramming. He was hired hundreds of times by aggrieved parents. Patrick's methods began with a kidnapping, followed by physical restraint and forced counseling.

Like Patrick, many deprogrammers did not have any formal counseling credentials, and so the process itself had much more in common with the Chinese brainwashing than it had with any legitimate psychotherapy. They would begin not by breaking down the subject's identity, but that of the cult leader, eventually getting the subject to recognize the reasons the cult leader has been rejected by moral society. At last the subject would see himself as allied with the deprogrammers against the cult leader.

Deprogramming was quite sticky from a legal standpoint. Obviously, kidnapping and false imprisonment are illegal, and many deprogrammers were convicted, including Patrick himself. The American Civil Liberties Union often took up the cases of deprogramming subjects, pointing out that they had joined the cults on their own free will, and it was the deprogrammers who were the only criminals involved. However, courts often found that parents and their hired deprogrammers were acting properly, since they believed (rightly or wrongly) that their children were in immediate danger. Usually the children were legally adults, and so police couldn't get involved with their disappearances; parents were on their own.

But as it turns out, deprogramming was never really necessary in the first place. The majority of research into actual cases of brainwashing has found that the effect is temporary at its best, and completely ineffective at its worst. Sociologists note that no cults have ever made any significant progress, thus demonstrating that any brainwashing techniques they may try to employ are not very effective. According to census numbers and the religions that people report, even the most aggressive mind control recruiters, like Scientology, grossly overstate their membership. In 1961, two books came out by prominent researchers who had studied the effects of the Chinese brainwashing on American POWs. Psychiatrist Robert J. Lifton *(Thought Reform and the Psychology of Totalism: A Study of "Brainwashing" in China)* and psychologist Edgar Schein *(Coercive Persuasion: A socio-psychological analysis of the "brainwashing" of American civilian prisoners by the Chinese Communists)* both found that most of the POWs had simply gone through the motions of saying and acting the way their captors wanted as a way to avoid punishment and obtain the rewards. Those few who actually did adopt the communist beliefs held them for only a very short time, and quickly reverted to their previous belief systems once they left captivity. Many deprogramming subjects turned out to have also been pretending to go along with the deprogramming, and returned to their cults shortly afterwards.

The U.S. court system agrees, now ruling claims of brainwashing as inadmissible, similar to lie detector results. According to what the legal system calls the Daubert standard, sciences are only ruled as admissible evidence when they're in line with the general scientific consensus on the subject, that the science has rigorous peer review, and that it has a known or discoverable error rate. In the case of brainwashing, this consensus comes from the American Psychological Association, which in 1987, withdrew a report they'd commissioned called the *APA Task Force on Deceptive and Indirect Techniques of Persuasion and Control.* Their withdrawal was based on their finding that the existence of brainwashing was not based on scientific and methodological rigor. They didn't say that it doesn't exist; they said there is insufficient evidence to conclude that it does.

Consequently, when convicted Beltway sniper Lee Boyd Malvo tried the "insanity by brainwashing" defense in 2003, it

didn't work. He would have had to prove that he'd been sufficiently dissociated from what he was doing that he had no conscious intention of killing the victims at the time he pulled the
trigger. Malvo had undoubtedly been encouraged and indoctri-
nated by his fellow sniper, the older John Allen Muhammad,
but the court found that his actions could not be excused. This
finding endorses Patty Hearst's conviction, as it can't plausibly
be argued that she had no conscious intention of robbing the
bank, when she had numerous opportunities to walk away.

So to sum it all up, brainwashing is probably not something
you need to be too worried about. Although it's been tried and
taken very seriously, the evidence of its effectiveness simply isn't
there. This makes deprogramming the solution to a problem
that doesn't exist. Deprogramming is really just brainwashing
under another name, and equally ineffective. Their midnight
abductions may have been spectacular, and their results impressive; but when you consider that their subjects were just ordinary young people who had made conscious decisions to try
some new far-out lifestyle, the value of understanding the real
science becomes clear. Of brainwashing and deprogramming,
you should always be skeptical.

REFERENCES & FURTHER READING

Barker, E. *The Making of a Moonie: Choice or Brainwashing?* New
York: Blackwell, 1984.

Flora, C. "The Brainwashing Defense." *Psychology Today.* Sussex
Publishers, LLC, 9 Dec. 2003. Web. 29 Sep. 2011.
<http://www.psychologytoday.com/articles/200312/the-
brainwashing-defense>

Hassan, S. *Combatting Cult Mind Control.* Rochester: Park Street
Press, 1988.

Lifton, R. *Thought Reform and the Psychology of Totalism: A Study of
"Brainwashing" in China.* New York: Norton, 1961.

Patrick, T., Dulack, T. *Let Our Children Go!* New York: Dutton,
1976.

Schein, E. *Coercive Persuasion: A Socio-Psychological Analysis of the "Brainwashing" of American Civilian Prisoners by the Chinese Communists.* New York: Norton, 1961.

25. NOAH'S ARK: SEA TRIALS

Could a wooden vessel like Noah's Ark actually have been made seaworthy?

Today we're going to have a bit of fun and shine the light of science on an ancient story. It is said that a gigantic wooden ship once carried a family and two of every kind of animal to safety, when the entire world was flooded. Noah's Ark sailed for five months, and then rested aground, sheltering its multitudinous crew for more than a year.

The elephant in the room here is that it's virtually impossible to write about this subject without having it sound like an attack on Christianity. I argue that it's not at all; the majority of Christians, when you combine the numerous denominations, don't insist that the Noah story is a literal true account. And, as has been pointed out many times, the Bible is hardly the only place where various versions of the Noah story are found. The most famous parallel, of course, is the Epic of Gilgamesh, wherein one of the many Babylonian gods charged the man Utnapishtim to build an ark, in a story that parallels Noah's in all the major details and most of the minor ones. It is perfectly plausible that all such stories stem from an actual event, the details of which are lost to history, but that might well account for the stories we have today of a boat and a flood. But regardless, in this chapter I'm not going to address any issues of faith, but only of science. We want to look at the engineering plausibility of Noah's great ship.

Noah's Ark was a great rectangular box of gopherwood, or perhaps some combination of other woods colloquially referred to as gopherwood. Its dimensions are given as 137 meters long,

23 meters wide, and 14 meters high. This is very, very big; it would have been the longest wooden ship ever built. These dimensions rank it as one of history's greatest engineering achievements; but they also mark the start of our sea trials, our test of whether or not it's possible for this ship to have ever sailed, or indeed, been built at all.

Would it have been possible to find enough material to build Noah's Ark? When another early supership was built, the *Great Michael* (completed in Scotland in 1511) it was said to have consumed "all the woods of Fife". Fife was a county in Scotland famous for its shipbuilding. The *Great Michael's* tim-

ber had to be purchased and imported not only from other parts of Scotland, but also from France, the Baltic Sea, and from a large number of cargo ships from Norway. Yet at 73 meters, she was only about half the length of Noah's Ark. Clearly a ship twice the length of the *Great Michael*, and larger in all other dimensions, would have required many times as much tim-

Michael ("Great Michael"), 1511

ber. It's never been clearly stated exactly where Noah's Ark is said to have been built, but it would have been somewhere in Mesopotamia, probably along either the Tigris or Euphrates rivers. This area is now Iraq, which has never been known for its abundance of shipbuilding timber.

In 2003, a doctoral candidate at the Georgia Institute of Technology, Jose Solis, created a proposal to build the Ark for Noah based on sound naval architecture. He proposed a dead weight — the weight of the wooden structure alone minus cargo and ballast — as 3,676 tons. Fully loaded, it would have displaced 13,000 tons, as compared to the *Great Michael's* 1,000 that consumed "all the wood of Fife". Where would all that wood have come from? In his proposal, Solis simply skipped this detail, and assumed the wood was commercially available at a cost of $16,472,040 in 2003 dollars. Tens of thousands of massive timber-quality trees would have to have been imported into the middle of what's now Iraq. Did Noah have the resources to import from France, Norway, or anywhere else?

But if the Ark did get built, it would be necessary to overcome its extraordinary fragility. If you pick up a toy Hot Wheels car, you can squeeze it as hard as you want but you can't break it. However, if you were a giant and reached down to pick up a normal passenger car, your fingers would crush it before creating sufficient friction to lift it. If you even lifted it by one corner, you would warp its structure noticeably. When we extend this to even larger vessels, their fragility is magnified. Recall that when the *Titanic* sank, that massive steel structure tore completely in half simply because one end was heavier than the other. Just that difference in weight was sufficient to tear open many decks of reinforced steel that had been engineered to the day's toughest standards. Were *Titanic* a wooden box instead of rigid steel, you (as a giant) could destroy it just by swishing your finger in the water next to it.

Allow me to explain. What's known as the square-cube law is pretty familiar: increase an object's dimensions, and its surface area increases by the square of the multiplier, and its weight increases by the cube of the multiplier. But one extension of this law is less familiar. When we scale up an object — take a wooden structural beam as an example — the strength of the beam does not increase as fast as its weight. Applied mechanics and material sciences give us all the tools we need to compute this. In summary, the tensile strength of a beam is a function of its moment and its section modulus. No need to go into the complicated details here — you can look up beam theory on Wikipedia if you want to learn the equations. Scale up a simple wooden beam large enough, the weight will exceed its strength, and it will break from its own weight alone. Scaled up to the immense size of Noah's Ark, a stout wooden box would be unspeakably fragile.

If there were even the gentlest of currents, sufficient pressure would be put on the hull to open its seams. Currents are not a complete, perfectly even flow. They consist of eddies and slow-moving turbulence. This puts uneven pressure on the hull, and Noah's Ark would bend with those eddies like a snake. Even if the water itself was perfectly still, wind would expose the flat-sided Ark's tremendous windage, exerting a shearing force that might well crumple it.

Whether a wooden ship the size of Noah's Ark could be made seaworthy is in grave doubt. At 137 meters (450 feet), Noah's Ark would be the largest wooden vessel ever confirmed to have been built. In recorded history, some dozen or so wooden ships have been constructed over 90 meters; few have been successful. Even so, these wooden ships had a great advantage over Noah's Ark: their curved hull shapes. Stress loads are distributed much more efficiently over three dimensionally curved surfaces than they are over flat surfaces. But even with this advantage, real-world large wooden ships have had severe problems. The sailing ships the 100-meter *Wyoming* (sunk in 1924) and 99 meter *Santiago* (sunk in 1918) were so large that they flexed in the water, opening up seams in the hull and leaking. The 102-meter British warships *HMS Orlando* and *HMS Mersey* had such bad structural problems that they were scrapped in 1871 and 1875 after only a few years in service. Most of the largest wooden ships were, like Noah's Ark, unpowered barges. Yet even those built in modern times, such as the 103 meter *Pretoria* in 1901, required substantial amounts of steel reinforcement; and even then needed steam-powered pumps to fight the constant flex-induced leaking.

Even in the world of legend, only two other ships are said to have approached the size claimed for Noah's Ark. One was the Greek trireme *Tessarakonteres* at 127 meters, the length and existence of which is known only by the accounts of Plutarch and Athenaeus. Plutarch said of her:

> *But this ship was merely for show; and since she differed little from a stationary edifice on land, being meant for exhibition and not for use, she was moved only with difficulty and danger.*

The other example is the largest of the Chinese treasure ships built by the admiral Zheng He in the 15th century, matching Noah's 137 meters, but only in the highest estimates. Many believe the biggest ships Zheng took with him on his seven voyages were no bigger than half that size, and moreover, that they remained behind in rivers and were not suitably seaworthy for ocean travels.

The long and the short of it — no pun intended — is that there's no precedent for a wooden ship the size of Noah's Ark being seaworthy, and plenty of naval engineering experience telling us that it wouldn't be expected to work. Even if pumps had been installed and all hands worked round the clock pumping, the Ark certainly would have leaked catastrophically, filled with water, and capsized.

There's another elephant in the room, too, that is necessary to address. Many of the problems with the Noah story are often answered, by those who regard it as a literal true account, with a special pleading. A special pleading is when any question is answered with "It was done by a higher power that you and I are not qualified to understand or question." Obviously, every point that science might raise regarding the Noah story can be fully answered with a special pleading. *Superman, Underdog,* and *The Jetsons* can shown to be literal true accounts if we allow special pleadings to be admissible. If the special pleading of divine intervention did indeed come into play during the Great Flood, then it was the most flagrant Rube Goldberg solution I've ever heard of. If divine intervention was needed to give Noah knowledge of how to build the Ark, or to provide the wood for its construction; then why not just provide an already-completed ark? Why bring the animals on board to be fed for a year or more, when divine intervention could have provided them an island? For that matter, why have the entire flood at all, when divine intervention could have simply struck down the evil humans with a plague? Why construct this most elaborate of all disaster and survival scenarios, some part of which was dependent on divine intervention; when divine intervention could have easily made the entire ordeal unnecessary? Special pleadings dismiss the true sciences that have allowed us to build real ships and conquer the world. Looking at the reality of what's possible and how things are done is always more interesting than imagining what's possible when anything is possible.

References & Further Reading

Burns, T. "Doctoral student weighs the cost, structure of a famous ship." *The Whistle.* Georgia Institute of Technology, 19 Apr. 2004. Web. 8 Oct. 2011.
<http://www.whistle.gatech.edu/archives/04/apr/19/ark.shtml>

Church, S. "Zheng He: An Investigation into the Plausibility of 450-foot Treasure Ships." *Monumenta Serica.* 1 Jan. 2005, Volume 53: 1-43.

Finlay, R. "How Not to (Re)Write World History: Gavin Menzies and the Chinese Discovery of America." *Journal of World History.* 1 Jun. 2004, 2004.

MacDougall, N. *Scotland & War: AD 79 - 1918.* Savage: Barnes & Noble, 1991. 36-57.

Plutarch. *Demetrius Poliorcetes and Antonius. Selections.* Cambridge: Cambridge University Press, 1988. Chapter 43.

Whitmer, E. "Elementary Bernoulli-Euler Beam Theory." *MIT Unified Engineering Course Notes.* 1 Jan. 1991, Section 5: 115-164.

26. FINDING SHAKESPEARE

Was William Shakespeare the author of his own works,
or was he merely someone else's pseudonym?

As any historical text on the subject will tell you, William Shakespeare was born in Stratford-upon-Avon in England, probably in 1564, and died in his hometown in 1616. He married and raised three children, and had a successful career in theater in London; so successful, in fact, that he was memorialized in effigy in the cathedral in Stratford-upon-Avon and even in London's own Westminster Abbey. The reason for this success is what you'll find in virtually any literary history: that Shakespeare is widely considered to be the finest playwright who ever lived, in any language, in any era. The "Bard of Avon", as he came to be known during his life, overcame his ordinary middle-class station and relative lack of formal education to compete with the finest noble playwrights of the day, and eventually trumped them all.

His is a fine story itself; a tale of great personal accomplishment, where talent and perseverance won out over the snobby, high-class competitors in the Elizabethan court. And during his life, few denied him the credit he was due.

But as we see so often with many of the subjects we cover on *Skeptoid,* an event that is unremarkable at the time it occurs is often magnified and misrepresented decades (or centuries) later into an extraordinary mystery, by people who weren't there and have no firsthand knowledge. After his death, Shakespeare's fame only grew (though he was not yet considered as great as he now is), and continued to grow for well over a century. In fact it wasn't until the late 1800s, generations after

Shakespeare died, that a few fringe authors began raising the specter of doubt over whether the Bard was indeed the author of the works attributed to him; or whether he was perhaps only the public face of the true author whose identity remains a secret.

The theories are many and varied. In some, an aristocrat is said to have written Shakespeare's works under a pseudonym because it was inappropriate for men of their rank to engage in commercial writing endeavors. In others, more educated authors are believed to have written the works because of the many references to high society and the royal court that they believe Shakespeare, who was of the merchant class, would not have known about. Still others raise the possibility that Shakespeare the actor from Stratford-upon-Avon may have simply coincidentally have had the same name as Shakespeare the author.

But the principal pieces of evidence against William Shakespeare are in three parts: First, that there is evidence he was illiterate. Second, that there are gaps in his documented history as a real living person. And third, the questions of his education and social status.

Shakespeare's presumed illiteracy is supported by scant evidence. There are only seven surviving signatures of his, and oddly, some are spelled differently from one another, and all appear to be nearly illegible scrawls. No signatures at all survive from Shakespeare's parents or from two of his children, except for marks used in place of signatures on legal documents. Marks were typically used by the illiterate. But all this is evidence of is that he was at least as literate as anyone in his family. The style of handwriting common in Shakespeare's time, known as secretary hand, often incorporated breviograms, shortened forms of words. Whether the various spellings of Shakespeare's signatures were breviograms or the result of either illiteracy or simple laziness, can't be known. It does not prove that Shakespeare the man was different from Shakespeare the author.

Shakespeare the man also left no correspondence. However, it turns out that this is the rule with authors of the day, not the exception. There are no original documents at all left by Christopher Marlowe, for one example, who was arguably even more famous than Shakespeare at the time. Not only is there a lack of documents by him, there is a lack of documents about him; where he went, what he did. Most scholars agree that this is not terribly unexpected. In his day he was a playwright and actor in London, a city where there were many playwrights and actors. Shakespeare's fame largely came in later centuries; during his lifetime, he was no more expected to have his activities documented than anyone else. Marlowe is well documented largely because he was often in trouble with the law and was also murdered. Their contemporary Ben Jonson's history is well known, as he was employed by royals and was the first playwright to receive an annual salary from the crown. Like most of his contemporaries, Shakespeare's private life appears to have been relatively unremarkable.

The four main nominees put forth by the anti-Stratfordians — those who doubt William Shakespeare's authorship — are Christopher Marlowe, Francis Bacon, William Stanley the 6th Earl of Derby, and Edward de Vere the 17th Earl of Oxford. Let's take a quick look at each. We won't go into great detail, because none of these are taken seriously by the overwhelming number of Shakespeare academics, but feel free to look more deeply into each on your own.

Marlovian theory states that the famous playwright Christopher Marlowe faked his own murder in 1593, and then wrote Shakespeare's plays. The timing works out pretty well to support this, but that's about all. Though there are questions about Marlowe's death, there's no credible evidence that he didn't actually die. Supporters have often claimed that the style of writing between the two men is too similar to be coincidental. There's little doubt that Marlowe was one of Shakespeare's main influences, but we'll talk more about the style similarities in a moment.

Baconian theory claims that the eminent Sir Francis Bacon wrote Shakespeare's works under a pseudonym to protect his reputation as a man of high standing. Throughout the 1800s, Bacon was the leading candidate among the anti-Stratfordians, many of whom claimed to find codes and ciphers within Shakespeare's works, wherein Bacon was trying to drop hints of his true identity. But searches for such clues are really just ex post facto rationalizations, and have never held up to serious scrutiny.

Derbyite theory points to the 6th Earl of Derby as the ringleader of a group of authors who collaborated to produce Shakespeare's works. The best evidence for this theory is no better than some of Shakespeare's plays seemed to parallel events in Derby's life; and the only suggestion for a motive that I've found is that Derby's family was a possible claimant to the English throne, so he felt it best to avoid politics and devote himself to cultural pursuits. Beyond a few nineteenth and early twentieth century authors, few have taken Derbyite theory seriously.

This brings us to the claim that has survived all the others: Oxfordian theory, promoting Edward de Vere, the 17th Earl of Oxford, as the true author. It was first proposed some 300 years after Shakespeare's death, by authors noting circumstantial evidence, such as the Oxfords' association with the theater and wealthy patrons. Some did speak well of de Vere's skill at poetry, but others have pointed out that reviewers might well be expected to give exaggerated praise to wealthy and respected men. Oxfordian theory has survived long enough that the 2011 movie *Anonymous* gives it as fact, much in the way that *Amadeus* promoted the untrue legend that Antonio Salieri murdered Wolfgang Mozart. It's well known that de Vere's family did participate in the publication of Shakespeare's works after his death, called the *First Folio*. But as evidence, this is only as convincing as suggesting that Stephen King's publisher must be the true author of his books. There's no evidence; only a supposition.

And this is really the best way to encompass all of the evidence that someone other than Shakespeare wrote his works: supposition. Logically, it's the same as 9/11 conspiracy theories. Look at some event or relationship a certain way, and it's always possible to find some circumstance to be consistent with just about any invented theory you like. But "consistent with" does not mean "evidence of".

There was one lesson that, in particular, has always stuck with me from my days studying screenwriting at UCLA. I don't know the actual quote or its source, but the story told by one of my instructors was that composing a play is like building a Frankenstein monster. He can teach you all the anatomy;

what connects to what, how to implant the brain, how all the pieces go together, what are all the dramatic elements of a good story. But the one thing he can't teach is the bolt of lightning that makes it all come to life. That lightning is the native genius of the author. You either have it or you don't. There are composers who study their entire lives and go to the best schools, but they will never be Mozart. Edward de Vere may have gone to Oxford and Cambridge and have been as well practiced as any poet on the planet, and William Shakespeare may have come from a small town with only a basic public education; but that spark of lightning was born to one and not to the other. Study and practice can improve your work, but it cannot create true genius. The work attributed to William Shakespeare is the product of true genius, not the product of education and social rank.

But let us not speculate. It turns out that technology finally did evolve to the point where we've been able to conclusively exclude all of these nominees, Edward de Vere the Earl of Oxford included, as having written Shakespeare's works. Computational stylistics is a branch of computer science in which a "literary fingerprint" can be determined for any author, based on computational analysis of his writing. As detailed in their 2009 book, *Shakespeare, Computers, and the Mystery of Authorship*, professors Arthur Kinney and Hugh Craig proved during their 2006 research at the University of Massachusetts Amherst that a single unique individual was the author of all the Shakespeare works, and nobody else. These computational techniques also made it possible to determine which plays influenced which later authors, and many other subtleties that escape conventional study of the texts. Hollywood movies to the contrary, we now know for a fact that neither de Vere of Oxford nor anyone else whose literary fingerprint differs deserves credit for William Shakespeare's life's work.

And neither was the Bard of Avon likely to take such charges lightly. A stone slab covers his remains at the Holy

Trinity Church in Stratford-upon-Avon, and on it is carved the following warning:

> *Good frend for Iesvs sake forbeare,*
> *To digg the dvst encloased heare.*
> *Bleste be ye man yt spares thes stones,*
> *And cvrst be he yt moves my bones.*

References & Further Reading

Callahan, P. "Computerized Analysis Helps Researchers Define Shakespeare's Work Using "Literary Fingerprint"." *Office of News and Media Relations.* University of Massachusetts Amherst, 27 Sep. 2006. Web. 10 Oct. 2011.
<http://www.umass.edu/newsoffice/storyarchive/articles/39476.php>

Joyrich, R. "Shakespeare Oxford Society." *Dedicated to Researching and Honoring the True Bard.* Shakespeare Oxford Society, 31 Dec. 1996. Web. 13 Oct. 2011. <http://www.shakespeare-oxford.com/>

Kinney, A., Craig, H. *Shakespeare, Computers, and the Mystery of Authorship.* New York: Cambridge University Press, 2009.

McCrea, S. *The Case for Shakespeare: The End of the Authorship Question.* Westport: Praeger, 2004.

McMichael, G., Glenn, E. *Shakespeare and his Rivals: A Casebook on the Authorship Controversy.* New York: Odyssey Press, 1962.

Nicholl, C. "Yes, Shakespeare Wrote Shakespeare." *The Times Literary Supplement.* 20 Dec. 2010, Number 5586: 3-4.

Shermer, M. "Skeptic's Take on the Life and Argued Works of Shakespeare." *Scientific American.* Scientific American, Inc., 31 Jul. 2009. Web. 12 Oct. 2011.
<http://www.scientificamerican.com/article.cfm?id=skeptics-take-on-the-life>

27. THE SCIENCE OF VOTING

Are democratic elections actually as free and fair as we think they are?

In the 1969 film *Putney Swope*, members of the board of executives were prohibited from voting for themselves, so they all voted for the one board member they were sure nobody else would vote for. Ergo, this free, democratic election produced a chairman that no voter wanted.

In a perfect democracy, everyone gets an equal opportunity to vote, and equal representation. Therefore, we hold elections to let everyone have their say, to either vote representatives into office, or to enact certain laws. It's a fine idea, and most countries do their level best to implement such systems. Some voters take advantage of it, and some choose apathy and don't vote. Some try to anticipate what other voters might do, and cast a vote in an unexpected direction not to vote for a candidate, but to affect another candidate's chances. This is what went wrong in *Putney Swope:* each voter cast a throwaway vote hoping to improve his own chances. In most elections, everyone has the right to do any of these things; the election is theirs, and theirs to decide. But what many of them might not know is that virtually any electoral process is flawed. Some outcomes are surprising. There are a number of different circumstances in which the candidate most desired does not win.

Democratic voting is only simple if there are just two candidates, or if it's a Yes or No vote. In those cases, any attempt to vote tactically or to create a voting bloc — casting votes that don't represent your preference — work against you. What we're talking about today are elections where there are three or more candidates. And the idea that all the various systems for running such elections are flawed (subject to results that do not represent the group's preference) is not just a whim or a crazy opinion of mine. It's proven by Arrow's Impossibility Theorem, named for the economist Kenneth Arrow, winner of the 1972 Nobel Prize in economics and the 2004 National Medal of Science. He proved it in 1951 with his Ph.D. thesis at Columbia University.

Arrow's theorem can be simplified into one clear statement: that no fair voting system exists when there are three or more candidates. To this I ask: What do you mean by fair? That's the key to Arrow's theorem. It holds true, depending on a rigid definition of fair that must satisfy three criteria:

1. If every individual prefers X to Y, then the group prefers X to Y.

2. If every voter's preference of X over Y stays the same, then the group's preference of X to Y stays the same, even if other preferences change: such as Y to Z, or Z to X.

3. There can be no dictator, as Arrow called him; a single voter with the power to dictate the group's preference.

Arrow's theorem applies to election systems that require voters to rank the candidates. This is the case with most voting systems worldwide. Typically, when you vote, you mark an X in the box for one candidate. That's a ranking; you've ranked that candidate first. Arrow's theorem applies to these simple ranking systems, but its richest mathematical complexities come from systems with three or more candidates and the voters rank all candidates in order of preference. This isn't used in many real-world elections, but it's the theoretical basis for social choice theory.

The ideal outcome in any election is to choose what's called the Condorcet winner. A Condorcet winner, named after the French mathematician and political scientist, is the candidate who would beat all other candidates in a simple two-man majority race. There isn't always a Condorcet winner in every election, but there usually is. The most common voting system is plurality voting, where the candidate with the largest number of votes wins. However, there are numerous situations in which the winner of a plurality vote is not the Condorcet winner who should have been elected. This is most often seen in a vote-splitting situation, where there are two similar candidates and one oddball candidate. One of the similar candidates is often the Condorcet winner; but because of their similarity, party votes are often split between them, and the oddball candidate wins. This is the most obvious failure of election systems, and it's exactly what Kenneth Arrows was talking about.

Experiments conducted by Don Saari, professor of mathematics and economics at the University of California Irvine, highlighted these failures of existing plurality systems. In one test, he had voters rank beer, wine, and milk in their order of preference. They all did so. But then there are different things you can do with those results. Saari made it a simple plurality election, giving a vote to each first-ranked item, and milk won; but he found that a majority of the voters would have preferred either beer or wine to it. The net result was that Saari proved that some system could be designed to select any desired outcome you want, given the same set of votes to work from.

This inherent tendency for voting to fail is called the voting paradox, also described by Condorcet. Yet, it's the system that virtually all nations rely upon for most or all of their elections. Something's broken somewhere.

Perhaps the most interesting solution to the election problem is called lottery voting, or the random ballot. In this type of election, every voter would cast his vote; but the winner would be determined not by a count, but by a single ballot selected at random from the total. Although at first it sounds outrageous, it actually has serious benefits. First, it renders irrelevant all attempts at strategic voting. Each voter knows that if his ballot is selected, the candidate he wrote will be the winner. There is no upside at all to trying to split the vote, or to vote for your second choice knowing he has a better chance of winning. Lottery voting is one of the only systems in which the voter's best strategy is always to vote for his most preferred candidate.

In conventional election systems, oddball candidates that may only have 5% of the vote do not have any realistic chance of winning the election, particularly in an Electoral College system like United States Presidential elections where they will always receive 0 electoral votes. But in lottery voting, these candidates would actually have a 5% chance of winning.

Another interesting alternative is range voting. The best example of range voting is the scoring at athletic events where each judge holds up a scorecard. Every voter gives a number to each candidate, say 0 to 10, and each candidate's total scores are averaged. Range voting has a lot of benefits that are attractive to voters. You can give everyone a 0, or you can give everyone a 10. You can express your thoughts about one or more candidates without wasting your vote, and still be able to give a high score to your preferred candidate. If there's a candidate you're not familiar with, you don't have to give any number to them, and you will not affect their average.

Significantly, every candidate has a realistic chance of winning in range voting. Voters don't need to vote for a potentially less-favored candidate simply because he has a better chance to win than their favorite candidate. For this reason, range voting is often touted by supporters of minority political parties. In fact, range voting often confers upon minority candidates a benefit called the nursery effect. In experiments, it turns out that people will give either an honest or a median vote to candidates who are relatively unknown; while at the same time being more likely to give the lowest votes to an opposing-party candidate that they don't like, but who may be far more experienced than the unknown candidate. Thus range voting can give bestow unearned support onto minority candidates.

But some in the major political parties advocate for it as well. A favorite example of US Democrats is the 2000 Presidential election, won by the Republicans after recounts of the Florida vote. Had range voting been in effect, Florida Democrats could have expressed their support for both Al Gore and Ralph Nader, giving Nader the Florida support he'd earned but without taking any crucial support away from Gore, that many say cost him the election. On the flip side, Republicans who gave their vote to Ross Perot in 1992 need not have been forced to choose between him and George H.W. Bush: They could have given both high scores, and Bush probably would have beaten Bill Clinton. Range voting eliminates both strategic voting and vote splitting, and in simulations, always elects the Condorcet winner.

But going back to Arrow's theorem, this should be an impossibility. Range voting appears to be fair to all the voters and to all the candidates. Therefore, it would seem to be a violation of Arrow's theorem. But it turns out that it's not. The reason is because range voting is a numerical rating system, whereas Arrow's theorem applies to rank orderings. When range voting, you can give the same score to multiple candidates; you do not have to rank them in order. There is no requirement to always prefer one candidate over the other, so it's a different type of system than those to which Arrow's theorem applies.

A variant of range voting is called approval voting, where you vote either for or against each candidate, but you can vote for as many as you like; even all or none. It's basically range voting with only two choices, 0 or 1. Approval voting also avoids the pitfalls of Arrow's theorem because it does not require ranking. It's simpler than range voting; and in both real-world and simulation examples, it selects the Condorcet winner virtually every time that one exists, in contrast to plurality voting which fails frequently.

Repairing our election methodology is, undoubtedly, a process fraught with political pitfalls, but I'll leave those to the political bloggers to hash out. The science says to listen a little more closely to Kenneth Arrow, and abandon rank-based voting systems. Change to any system that allows voters to express their preferences without requiring them to force-rank one candidate as the best, and the winner most approved of by the group as a whole will be elected far more often than not.

References & Further Reading

Amar, A. "Choosing Representatives by Lottery Voting." *Yale Law Journal.* 1 Jun. 1984, Volume 93, Number 1283: 1-34.

Aron, J. "Mathematicians Weigh In on UK Voting Debate." *New Scientist.* Reed Business Information Ltd., 27 Apr. 2011. Web. 14 Oct. 2011.
<http://www.newscientist.com/blogs/shortsharpscience/2011/04/mathematicians-weigh-in-on-uk.html>

de Caritat, N., marquis de Condorcet. *Essay on the Application of Analysis to the Probability of Majority Decisions.* Paris: Institut de France, 1785.

Saari, D. *Chaotic Elections! A Mathematician Looks at Voting.* Providence: American Mathematical Society, 2001.

Smith, W., Kok, J. "Range Voting." *RangeVoting.org.* The Center for Range Voting, 1 Aug. 2005. Web. 19 Oct. 2011.
<http://rangevoting.org>

Stewart, I. "Electoral Dysfunction: Why Democracy Is Always Unfair." *New Scientist.* 28 Apr. 2010, Issue 2758: 28-31.

28. THE JERSEY DEVIL

This creature has been haunting New Jersey for nearly 300 years. From whence did it come?

For eight days in 1909, residents of New Jersey were terrorized by one of the strangest creatures in all of legend. Called the Jersey Devil, it stood a little shorter than a man, with the body of a serpent and the head of a horse. It had cloven hooves on its feet, its two arms were small and rarely used, and it had a devil's forked tail. Most notably, it flew on great leathery batlike wings, and had a horrifying screech. The newspapers were filled with dozens and dozens of accounts: the Jersey Devil killed chickens and dogs, flapped and screeched and chased residents, and left cloven footprints in the snow all throughout the region. Police officers, fire departments, and public officials were named in the reports. Given its comprehensive newspaper coverage, the Jersey Devil's 1909 scourge is perhaps the best documented of all cryptids.

Although 1909 was the Jersey Devil's heyday and by far its most active period, other scattered reports have trickled in ever since, and for more than a century and a half before. The history of the Jersey Devil's beginnings is widely available on the Internet and in print, and is often, in fact, disappointingly copy-and-pasted from one article to another. There are references to the earliest and best sources, but little effort seems to have been made to go back and actually dig them up. If we want to learn what the true cause of these reports was, we need to go back to the original sources, and avoid the recycled retellings published today. Here's what we know for a fact, from historical records.

Leeds Point is a small triangular protuberance into the swampy shallows of Great Bay just north of Atlantic City. The surrounding area is called the Pine Barrens, one million acres of dense pine trees where a disoriented traveler could quickly become lost. It was surveyed by, and subsequently granted to, a Mr. Daniel Leeds, an Englishman who had immigrated in 1678. He was best known as the publisher of *The American Almanack,* which he printed until his retirement in 1716, when it was taken over by his sons. Benjamin Franklin actually referred to Leeds by name in 1735 in his own *Poor Richard's Almanack,* and called him "the first author south of New York."

The popular tale you're likely to read holds that Leeds' wife, a Quaker named Deborah Smith, gave birth to her thirteenth son in 1735 in Estelville, some 30 km west of Leeds Point. Some say either Mother Leeds or a clergyman cursed the infant; some say it was born horribly deformed; some say it was born normally but quickly transformed into a monster who killed Mother Leeds and then escaped. In all likelihood, whatever the actual birth was, it seems that the poor Mother Leeds and her infant both died in childbirth.

But in looking at the historical sources, we soon find that this story is not possible. First, Daniel Leeds died in 1720, fifteen years before the fabled birth; second, Daniel Leeds was married to Ann Stacy; and when she died in childbirth, he married Dorothy Young, to whom he remained married until his death. There is no Deborah Smith the Quaker in his history.

Prof. Fred R. MacFadden, Jr. of Coppin State University in Baltimore (formerly Coppin State College) did substantial research on Leeds and also into the Jersey Devil's earliest mentions in print, and some of this was published in William McMahon's 1973 book *South Jersey Towns: History and Legend.* It turns out that the date of 1735 comes from the earliest print reference to a "devil" that he could find, and its location was given only as Burlington, which was the name of whole county or region at the time. There appears to be no contemporary sources connecting Daniel Leeds or either of his wives to a devilish character of any sort, and MacFadden himself was only able to speculate whether the 1735 Burlington mention of a "devil" may refer to the same beast popularized in 1909 and known today. Although newspapers of the 1800s did occasionally print the Mother Leeds story as given in the legend, we seem to have a total lack of factual basis to anchor it to any real history.

The character of Daniel Leeds the man may have something to do with the connection of the monster to the Leeds name. As an editor and local politician, he had friends and enemies. Most notoriously, Leeds served as deputy to Edward Hyde, Lord Cornbury, who served Queen Anne as colonial governor of New York and New Jersey. But in 1708 Lord Cornbury was recalled to England due to his unpopularity; and Leeds, unpopular by association, withdrew from public office and never again served in politics. As the deputy of a disgraced governor, he was wide open to criticism and ridicule. Even his religion was controversial, as a former Quaker turned Episcopalian. According to MacFadden, several of Leeds' children were mentally disabled. Putting all of these facts and allegations together, Daniel Leeds was a gigantic easy target for anyone who wanted a name upon which to hang a ghastly monster tale.

So well connected is the Jersey Devil's beginnings to Daniel Leeds, that in most early accounts it is called the Leeds Devil. In fact, a search of newspaper archives reveals that almost all of its references prior to 1909 are to the Leeds Devil, not the Jersey Devil. The namesakery bears the strong stench of politics. When Thomas Edison wanted to discredit the alternating current promoted by Westinghouse, he tried to frighten the public away from it by referring to electrocution as Westinghousing. In the 1800s when reports began to appear, Daniel Leeds was long dead; but his was a well-known name, and newly minted Americans were always happy to have a British loyalist at whom to throw mud.

By 1909 nobody cared about Daniel Leeds anymore, and the creature has mainly been known as the Jersey Devil ever since.

Investigator Joe Nickell has also found another source for the 1909 reports:

> *In January 1909 the monster was revived by a hoax. Displayed in a private museum in Philadelphia, the creature was actually a kangaroo outfitted with fake wings affixed by a harness. To make it leap at spectators when the curtain was drawn, a boy hidden at the rear of the cage prodded the unfortunate animal with a stick.*

The exact date of this exhibit is not clear, but there's at least one clue that indicates it was an attempt to capitalize on the existing Devil craze, rather than the spark that ignited it. MacFadden writes that Gloucester county archivist R.C. Archut tracked down a photograph of cloven footprints on a snowy porch, dated 1908, the beginning of the period during which the footprints were reported. Nickell's date of January 1909 for the hoax exhibit strongly suggests that the hoax did not initiate the 1909 cluster of sightings.

So what did start things off? We don't know, and at this late date, we probably never will. It's always better to admit that we don't know facts that we don't know, rather than to assert that the lack of strong, testable evidence means nothing happened at all.

In most cases like this, some of the sightings turn out to be mistaken identifications of everyday animals or something else. There is one decent candidate that's been put forward for the Jersey Devil, and that's the New Jersey species of the Sandhill Crane. It's a big slender bird; it can stand tall, and it can spread and flap its two-meter batlike wings. However in 1887, a J. A. Singley wrote in to the *Galveston Daily News* that in his experience, the Leeds Devil described as a "bugaboo bird" by correspondent Sam E. Hayes was probably just a barn owl:

> *The bird has an uncanny appearance generally, and this, aided by a lively imagination, has probably originated the blood-curdling story of the Leed's Devil.*

Cranes were also suggested as a likely cause of the 1909 footprints. In 1926, Alfred Heston wrote in *Jersey Waggon Jaunts* that residents of Salem called the 1909 creature the Ostrich Devil. I was not able to track down a copy of the Archut photograph, but MacFadden did note that cranes were often associated with the type of tracks reported. Moreover, there are birds that hop with both feet together for whom there is no rational foundation to be ruled out; and as we discussed when examining the similar case of the Devon, England footprints from 1855, there are lots of other common animals that might have been responsible for the footprints.

Of course, most of these suggestions for alternate explanations of individual bits and pieces of the Jersey Devil canon are really just supposition. There is no more evidence that a crane is responsible for any of the honestly reported incidents than there is for the Jersey Devil being the real Satanic offspring of Mother Leeds. We've been able to piece together enough of the original history to disprove parts of the commonly told version of the monster's genesis, but that says nothing about the sightings that have happened since that date.

Here's what I think is the best way to evaluate any given report of the Jersey Devil. Take the story of the cab driver who was fixing a flat tire in 1927, when according to him, the creature landed on the roof of his car and shook it violently, compelling him to flee the scene. We can't examine his car for forensic evidence; we can only speculate. Maybe he made it up. Maybe it was a person or large bird, and for some reason he misidentified it. Maybe it was a joke spread by his buddies at the cab station. Maybe it happened exactly as he reported. Maybe it never happened at all, and was invented by a reporter, or misinterpreted by some third party, or told and retold until the story morphed into this version. All seven of these are possible, as well as others; we don't know. And that's the only thing we can know for sure. Each of the dozens of such anecdotes is equally ambiguous. What we have is an interesting bit of folklore; what we do *not* have is any compelling reason to conclude there are any Jersey Devils outside of Newark's professional hockey arena.

REFERENCES & FURTHER READING

Beck, H. *Jersey Genesis.* New Brunswick: Rutgers University Press, 1963.

Heston, A. *Jersey Waggon Jaunts: New Stories of New Jersey.* Pleasantville: Atlantic County Historical Society, 1926. 269.

MacFadden, F. "Claws, Hoof, and Foot: The Devil's Tracks in Devon and New Jersey." *Free State Folklore.* 1 Apr. 1976, Volume 3, Number 2: 5-14.

McCrann, G. "Legend of the New Jersey Devil." *Jersey History.* The New Jersey Historical Society, 26 Oct. 2000. Web. 23 Oct. 2011. <http://www.jerseyhistory.org/legend_jerseydevil.html>

Mcgloy, J., Miller, R. *The Jersey Devil.* Wallingford: The Middle Atlantic Press, 1976. 45.

McMahon, W. *South Jersey Towns: History and Legend.* New Brunswick: Rutgers University Press, 1973. 210-213.

Nickell, J. "Jersey-Devil Expedition." *Investigative Briefs.* Center for Inquiry, 26 Aug. 2010. Web. 22 Oct. 2011. <http://www.centerforinquiry.net/blogs/entry/jersey-devil_expedition/>

Singley, J. "Leaves from Literature." *Galveston Daily News.* 31 Dec. 1887, Newspaper: 7.

29. Top 10 Worst Anti-Science Websites

My list of the worst offenders on the web in the promotion of scientific and factual misinformation.

The Internet is a dangerous place. It's full of resources, both good and bad; full of citations linking one to another, sometimes helpfully, sometimes not. Today we're going to point the skeptical eye at ten of the worst web sites in terms of quality of science information that they promote. To make this list, they not only need to have bad information, they also need to be popular enough to warrant our attention.

Many of these sites promote some particular ideology, but I want to be clear that that's not why they're here. Sites that make this list are only here because of the quality of the science information that they advocate.

As a measure of each site's popularity, I'm giving its ranking on Alexa.com as of this writing. Of course this changes over time, so I'm rounding them off to give a general idea of each site's traffic. Also, I'm giving its US traffic ranking, as these are English language sites and the worldwide rankings are skewed by sites in China, Russia, and the rest of the non-

English world. For a starting point of reference, Skeptoid.com's ranking is currently about 40,000, meaning that 40,000 web sites in the United States get more traffic than I do. And, compared to the number of web sites there are, that number is actually not half bad — but note how it compares to some of these sites promoting misinformation.

Let's begin at the bottom of our list of the worst offenders, with a site that nevertheless has staggering amounts of traffic:

10. HUFFINGTON POST

huffingtonpost.com
Alexa ranked #23
Google PageRank 8

The Huffington Post is arguably one of the heaviest trafficked news, opinion, and information sources on the Internet. Its many editors and 9,000 contributors produce content that runs the gamut and is generally decent, with one exception: medicine. HuffPo aggressively promotes worthless alternative medicine such as homeopathy, detoxification, and the thoroughly debunked vaccine-autism link. In 2009, Salon.com published a lengthy critique of HuffPo's unscientific (and often exactly wrong) health advice, subtitled *Why bogus treatments and crackpot medical theories dominate "The Internet Newspaper"*. HuffPo's tradition is neither new nor just a once-in-a-while thing.

Science journalists have repeatedly taken HuffPo to task for this, and repeatedly been rebuffed or not allowed to submit fact-based rebuttals. HuffPo's anti-science stance on health and medicine appears to be deliberately systematic and is unquestionably pervasive.

9. Conservapedia

> *conservapedia.com*
> *Alexa ranked #13,600*
> *Google PageRank 5*

Conservapedia was founded by Christian activist Andrew Schlafly as resource for homeschooled children, intended to counter what he saw as an anti-Christian bias in Wikipedia and science information in general. It is, in short, an encyclopedia that gives a Young Earth version of every article instead of the correct version. If you want to know about dinosaurs, geology, radiometric dating, the solar system, plate tectonics, or pretty much any other natural science, Conservapedia is your Number One resource to get the wrong answer. That it is intended specifically as a science resource for homeschooled children, who don't have the benefit of an accredited science teacher, is its main reason for making this list.

8. Cryptomundo

> *cryptomundo.com*
> *Alexa ranked #41,800*
> *Google PageRank 5*

Run by cryptozoologists Loren Coleman, Craig Woolheater, John Kirk, and Rick Noll, Cryptomundo promotes virtually every mythical beast as being a real living animal. Cryptozoology may be a fun and illustrious hobby for some, but its method of beginning with your desired conclusion and working backwards to find anecdotes that might support it is pretty much the opposite of the scientific method. Cryptomundo only ranks as #8 on our list because, let's face it, cryptozoology is not exactly the most harmful of pseudosciences. It's more of a weekend lark for enthusiasts of the strange.

Cryptomundo's forum moderators have something of a notorious reputation for editing comments posted by site visitors, and for deleting comments that express skeptical points of view. Some skeptical commenters have reported even being banned completely from the forums, not for spamming or trolling, but just being consistently skeptical.

7. 9/11 TRUTH.ORG

911truth.org
Alexa ranked #109,000
Google PageRank 5

The only reason this site has such a low traffic rating is that its field is saturated with competition. 9/11 Truth.org is only the largest of the many, many web sites who began with the idea that 9/11 was a false flag operation against American citizens staged by the American government, but unlike most others, it has stayed on topic. Even more than a decade after 9/11, 911 Truth.org still manages to find and post articles almost daily promising to reveal new evidence proving the conspiracy.

6. MERCOLA.COM

mercola.com
Alexa ranked #650
Google PageRank 6

The sales portal of alternate medicine author Joseph Mercola has received at least three warnings from the U.S. Food and Drug Administration to stop making illegal health claims about the efficacy of its products. A tireless promoter, Mercola has built his web site into probably the most lucrative seller of quack health products. But Mercola's web site is not wrong because it's lucrative; it's wrong because the vast majority of its merchandise has no proven medical value, yet virtually all of its product descriptions imply that they can improve the customer's health in some way. Today's Featured Products include:

> *Probiotics supplements that can "boost your body's defense against disease and aid your production of essential nutrients".*

and

> *Krill oil that provides "A healthy heart, Memory and learning support, Blood sugar health, Anti-aging, Healthy brain function and development, Cholesterol health, Healthy liver function, Boost for the immune system, Optimal skin health".*

At least Mercola.com usually includes the required statement (tucked way down at the bottom of the screen in a tiny font) that "These statements have not been evaluated by the Food and Drug Administration. This product is not intended to diagnose, treat, cure or prevent any disease." Presumably that's a result of all the regulatory action he's suffered.

5. Answers in Genesis

> *answersingenesis.org*
> *Alexa ranked #9,800*
> *Google PageRank 6*

Evangelical Christian web sites are a fine thing for those who roll that way, and most such sites do good charitable and social works. But a few stray from that mission, and Answers in Genesis is the leading example. Their "Statement of Faith" is, in their own words:

> *Scripture teaches a recent origin for man and the whole creation, spanning approximately 4,000 years from creation to Christ. The days in Genesis do not corre-*

> spond to geologic ages, but are six [6] consecutive twen-
> ty-four [24] hour days of creation. The Noachian
> Flood was a significant geological event and much (but
> not all) fossiliferous sediment originated at that time.

There's no way around it: This is not doing any kind of a service mission, this is unabashed promotion of scientific misinformation. Even the world's largest Christian organization, the Catholic Church, rejects Answers in Genesis' alternate-reality version of geology, biology, and virtually every other natural science. Worse, AiG provides a wide array of highly polished, very professionally written educational materials including study guides, online courses, and lesson plans for teachers. So far the American court system has done a pretty good job of keeping this stuff out of public schools, but their penetration into private schools and homeschools is only growing.

4. AUSTRALIAN VACCINATION NETWORK

avn.org.au
Alexa ranked #21,600 (in Australia)
Google PageRank 4

Founded by Australia's best known anti-vaccine activist, Meryl Dorey, this site earns its recognition by the sheer magnitude of scientific, regulatory, and ethical criticism it has received. The AVN really put itself on the map with its refusal to post a disclaimer clearly identifying itself as anti-vaccine, as ordered by Australia's Health Care Complaints Commission. It's had its license to accept charitable donations revoked for multiple violations of the Charitable Fundraising Act, and its anti-science stance earned it a spot on *Australian Doctor* magazine's "Top 50 Medical Scandals of the Past 50 Years". If I wanted, I could write an entire book just listing the violations, criticisms, complaints, investigations, and regulatory actions the AVN has been hit with.

Yet it persists, boasts thousands of members, and continues to significantly reduce levels of immunity to infectious disease within Australia.

3. Prison Planet / InfoWars

prisonplanet.com
Alexa ranked #2,000
Google PageRank 6

infowars.com
Alexa ranked #566
Google PageRank 6

There doesn't appear to be any clear difference between Prison Planet and InfoWars, the websites of conspiracy theorist Alex Jones. Both sites are heavily trafficked collections of articles predicting the takeover of the world by nebulous Illuminati in the form of governments, companies and industries. There's nothing wrong with being anti-government and anti-corporate; they're perfectly valid philosophies, if that's the way you roll. Alex Jones' sites are on this list for having almost daily made predictions of New World Order takeovers, global currencies, and mass executions for many years, none of which have ever come true; and for distorting virtually every aspect of modern society into evidence of some vague worldwide plot to control or kill law abiding citizens.

2. Age of Autism

ageofautism.com
Alexa ranked #33,500
Google PageRank 5

This website of investigative reporter Dan Olmsted promotes his own notions that autism is caused by mercury toxicity (contrary to what we've learned scientifically), that it is increasing dramatically at epidemic proportions, not just in counting methods but in actual incidence (contrary to what's been measured), and that it can be cured by holistic treatments, supplementation, hyperbaric oxygen therapy, removal of dental fillings, and bowel cleansing (contrary to all research done on these methods).

Web authors like Olmsted obviously must know that their writing is at variance with science based findings, so there must be some kind of cognitive dissonance going on, outright dishonesty, or perhaps even a belief in a global Big Pharma conspiracy of bad science.

Lest you think that fringe cranks like Olmsted have no influence and their sites can be dismissed, Age of Autism articles were cited in a 2006 U.S. House of Representatives bill to reinvestigate the thoroughly debunked link between mercury and autism — using taxpayer funds to challenge science-based medicine.

1. Natural News

naturalnews.com
Alexa ranked #1,000
Google PageRank 6

When Natural News began, it was basically the blog and sales portal of anti-pharmaceutical activist Mike Adams. His basic premise has always been the Big Pharma conspiracy, the idea that the medical industry secretly wants to keep everyone sick, and conspires with the food industry to make people unhealthy, all driven by a massive plot of greed to sell poisonous medicines. Adams appears to have become a protégé of Alex Jones, for he now writes on Natural News at least as many police state conspiracy articles as he does anti-science based medicine articles. They carry ads for each other on their sites as well.

Some examples of current articles on Natural News are:

New World Order: Implantable RFID chips capable of remotely killing non-compliant 'slaves' are here

Vaccines lower immunity

Fluoride means lower IQs and more mental retardation

and of course:

Jumping rope and 9/11 truth - how the sheeple have been trained to avoid unpopular truth about WTC 7

Natural News' misleading title — I see very little on the site that I would think to classify as "natural news" — and pretense of being a health resource has helped it to become an often cited and heavily read site. For its frighteningly large influence, and abysmal quality of information, it earns the #1 spot on this list.

References & Further Reading

Barrett, S. "FDA Orders Dr. Joseph Mercola to Stop Illegal Claims." *Quackwatch.* Stephen Barrett, MD, 26 May 2011. Web. 1 Nov. 2011. <http://www.quackwatch.org/11Ind/mercola.html>

Novella, S. "Mike Adams Takes On 'Skeptics'." *Neurologica.* New England Skeptical Society, 25 Jan. 2010. Web. 1 Nov. 2011. <http://theness.com/neurologicablog/index.php/mike-adams-takes-on-skeptics/>

Parikh, R. "The Huffington Post is Crazy about Your Health." *Salon.com.* Salon Media Group, 30 Jul. 2009. Web. 1 Nov. 2011. <http://www.salon.com/2009/07/30/huffington_post/singleton/>

Pehm, K. "Letter to AVN." *Health Care Complaints Commission.* New South Wales Government, 7 Jul. 2010. Web. 1 Nov. 2011. <http://www.stopmeryldorey.com/wp-content/uploads/2010/07/HCCC-Report.pdf>

Phelps, D. "The Anti-Museum: An overview and review of the Answers in Genesis Creation Museum." *Defending the Teaching of Evolution in Public Schools.* National Center for Science Education, 17 Oct. 2008. Web. 1 Nov. 2011. <http://ncse.com/creationism/general/anti-museum-overview-review-answers-genesis-creation-museum>

Zaitchik, A. "Meet Alex Jones: The Most Paranoid Man in America." *Rolling Stone Magazine.* 17 Mar. 2011, Issue 1199.

30. THE FATE OF FLETCHER CHRISTIAN

Did the leader of the Bounty mutineers die on Pitcairn Island, or did he eventually make it back to England?

The mutiny on the *Bounty* is perhaps the best known of all stories from the era of wooden ships. Fletcher Christian, the infamous officer responsible for the affair, is believed to have died on Pitcairn Island, where he and the other mutineers took refuge. Yet some say his death was faked, and he did in fact make it back to England. Today we'll point the skeptical eye at these stories, and see if we can learn for certain where Fletcher Christian made his final atonement.

The basic story of the *Bounty* is not only well known, it's well documented and not in any meaningful doubt. In 1789, the small British naval ship left the island of Tahiti with a cargo of breadfruit plants. Three weeks later, its discontented crew, led by sailing master Fletcher Christian, mutinied against Captain William Bligh. Bligh and the loyal crew members were set adrift in the *Bounty's* open launch, in which they ultimately made it to safety, and made knowledge of the mutiny public.

Christian and his crew of 24 — eighteen mutineers, four loyalists who couldn't fit in Bligh's launch, and two neutral men — sought refuge for several months in some of the neighboring islands, but upon finding the natives too unfriendly, they returned briefly to Tahiti. Sixteen of the men remained there, leaving only eight aboard the *Bounty;* barely enough to sail her. And so, one night when the mutineers' women and some other natives happened to be on board, they set sail unexpectedly, effectively kidnapping the Tahitians. And thus was the founding population of Pitcairn Island established: eight British sailors, six Tahitian men, eleven Tahitian women, and one baby. These events are known from the accounts of the sailors who remained on Tahiti, including the four loyalists, who were either captured by or rejoined the British navy when the ship *Pandora* was dispatched to find them.

From that point onwards, the fate of the *Bounty* is more thinly documented. Fletcher Christian took his crew to Pitcairn Island because he knew from the British charts that its position was not precisely known, so they'd have a fair chance of evading capture. When they arrived, the *Bounty* was scuttled, both to avoid advertising their presence and to prevent anyone from leaving the island and possibly raising the alarm. We know for a fact that the *Bounty* was sunk because its remains have been found. Without any reasonable doubt, Fletcher Christian left Tahiti aboard a ship that went to Pitcairn Island and nowhere else. No other ship of any nation reported encountering them en route.

One of the mutineers who elected to remain on Tahiti was Peter Heywood, a close friend of Christian's. Along with the others, Heywood was captured by the *Pandora* in 1791 and returned to England. He was court martialed and sentenced to hang; but his was a family of wealth and influence, and Heywood received a pardon. Heywood returned to service in the navy, rose through the ranks, and had a successful career as a captain. Heywood was to play a pivotal role in the theories of Christian's alleged return to England. It was reported in 1831 by Sir John Barrow, an acquaintance of Heywood's, who detailed the following account in his book *The Mutiny and Piratical Seizure of HMS Bounty:*

> *In Fore-street, Plymouth Dock, Captain Heywood found himself one day walking behind a man, whose shape had so much the appearance of Christian's, that he involuntarily quickened his pace. Both were walking very fast, and the rapid steps behind him having roused the stranger's attention, he suddenly turned his face, looked at Heywood, and immediately ran off. But the face was as much like Christian's as the back, and Heywood, exceedingly excited, ran also. Both ran as fast as they were able; but the stranger had the advantage, and, after making several short turns, disappeared.*

> *That Christian should be in England, Heywood considered as highly improbable, though not out of the scope of possibility; for at this time no account of him whatsoever had been received since they parted at Otaheite; at any rate the resemblance, the agitation, and the efforts of the stranger to elude him, were circumstances too strong not to make a deep impression on his mind. At the moment, his first thought was to set about making some further inquiries; but on recollection of the pain and trouble such a discovery must occasion him, he considered it more prudent to let the matter drop; but the circumstance was frequently called to his memory for the remainder of his life.*

Although Heywood's is the only reliably documented account of anyone actually encountering Fletcher Christian in England after the mutiny, there was already something of an urban legend at the time. Barrow also wrote:

> *About the years 1808 and 1809, a very general opinion was prevalent in the neighborhood of the lakes of Cumberland and Westmoreland, that Christian was*

*in that part of the country, and made frequent private
visits to an aunt who was living there.*

In 1797, eight years after news of the mutiny reached England, Samuel Taylor Coleridge wrote his poem *The Rime of the
Ancient Mariner,* which in some circles was believed to be
loosely based on the life of Fletcher Christian, prompting speculation that Christian must have been available for Coleridge to
interview. Coleridge's colleague William Wordsworth had been
a childhood classmate of Christian's. When Christian was tried
in absentia for the crime of piracy, he was defended by his
brother, the lawyer Edward Christian, and Wordsworth joined
in his defense. So well known (or at least well believed) was the
association of Coleridge and Wordsworth with Christian that
at least one author, C.S. Wilkinson, proposed in his book *The
Wake of the Bounty* that the two poets might have collaborated
to have somehow brought Christian back to England. No evidence of this has ever been offered, but it remains one of the
most popular rumors about Christian's fate.

Could the man Peter Heywood chased in Plymouth actually have been Fletcher Christian? For this to be possible, Christian would have had to have found some way of leaving Pitcairn
Island after his known arrival there in 1789, and no later than
when the island was visited by the American seal hunting ship
Topaz in 1808. On that day, three young men who appeared to
be Pacific Islanders paddled out to the *Topaz* in a Tahitian style
canoe, and astonished its captain, Mayhew Folger, with their
friendliness and perfect English. According to Folger's log-
book, the three young men invited him ashore to dine with the
man they called their "father", Aleck. Aleck turned out to be an
Englishman named Alexander Smith, and was the sole surviving Englishman on the island. Aleck identified himself as one
of the crew of the Bounty, and gave Folger the general facts.
He also explained that the six Tahitian men, whom they had
kept as slaves on the island, rose up and murdered all of other
mutineers, including Fletcher Christian. Aleck and the women
then managed to put all six of the Tahitians to death, leaving

them in the current situation. Folger was ultimately able to deliver this report to the British navy, along with his statement of Aleck's character:

> ...He Immediately went to work tilling the ground so that it now produces plenty for them all and the[re] he lives very comfortably as Commander in Chief of Pitcairn's Island, all the Children of the deceased mutineers Speak tolerable English, some of them are grown to the Size of men and women, and to do them Justice I think them a very humane and hospitable people, and whatever may have been the Errors or Crimes of Smith the Mutineer in times Back, he is at present in my opinion a worthy man and may be useful to Navigators who traverse this immense ocean, such the history of Christian and his associates.

Folger was satisfied that there were indeed no other Englishmen living on the island. Similar circumstances were discovered six years later in 1814 when two British ships, HMS *Briton* and *Tagus*, visited. This time the leader of the young men identified himself as Thursday October Christian, the 25-year-old son of Fletcher Christian. By Thursday's own account, his father had indeed been killed on the island. Captain Pipon of the *Tagus* wrote a detailed account of their days spent on the island, and of what they learned. Alexander Smith, it turned out, was a fake name, and Aleck was actually John Adams, an Able Seaman who denied having had any part in the mutiny (contrary to what had already been learned from Captain Bligh and his loyalists).

Perhaps having learned from Folger that there was no longer any great dragnet to catch the mutineers, Adams was much more forthcoming with Pipot than he had been with Folger. He showed the detailed log the islanders had kept all those years, and it included the true fates of the Englishmen and the Tahitians. Disputes over women, authority and slavery had torn the group apart, with murders having taken place on both sides. Fletcher Christian had been killed by two of the Tahitians on the island's bloodiest day in 1793 on which four of the Englishmen died. Christian was survived by his Tahitian wife Maimiti and three children: Thursday, Charles, and Mary Ann. Thursday, the oldest, was not quite three years old when his father was killed, so Adams (and one other who had died in the interim) were the only Englishmen any of them ever really knew. Maimiti witnessed her husband's death and later recounted it in great detail. This was the true history of Pitcairn Island's colonists according to all of them who were ever asked.

For Fletcher Christian to have been the man that Peter Heywood chased, he would have had to survive on the tiny Pitcairn Island undetected by his own family for 15 years, then sneak on board the *Topaz*, somehow persuade its captain and crew not to reveal his existence, then found his own way back to England (halfway around the world) within the year while the *Topaz* and its crew were held by Spanish authorities on Juan Fernandez Island for several months on an unrelated matter. Is that string of improbabilities really more likely than Heywood was simply mistaken about the identity of a man whose face he saw only once in a quick glance, and even then only in a secondhand report?

The escape of Fletcher Christian, or any other larger-than-life character from history, makes for a fine story, but not necessarily a true one.

REFERENCES & FURTHER READING

Alexander, C. *The Bounty: The True Story of the Mutiny on the Bounty.* New York: Viking, 2003.

Barrow, J. *The Eventful History of the Mutiny and Piratical Seizure of H.M.S. Bounty: Its Cause and Consequences.* London: John Murray, 1831. 309-310.

Curry, K. *New Letters of Robert Southey, Volume 1, 1792-1810.* New York: Columbia University Press, 1965. 519.

Pipon, P. "Capt. Pipon's Narrative of the Late Mutineers of H.M. Ship Bounty Settled on Pitcairn's Island in the South Seas; in Sept 1814." *Fateful Voyage.* James Galloway, 4 Apr. 2010. Web. 10 Nov. 2011.
<http://www.fatefulvoyage.com/pitcairn/pitcairnBPipon1814.html>

Wahlroos, S. *Mutiny and Romance in the South Seas: A Companion to the Bounty Adventure.* Topsfield: Salem House Publishers, 1989.

Wilkinson, C. *The Wake of the Bounty.* London: Cassell, 1953.

31. KOREAN FAN DEATH

An urban legend in Korea states that running an electric fan at night can kill you.

Today we're going to point the skeptical eye at a traditional belief from Korea, one at which many Westerners merely scoff. Many Koreans believe that sleeping in a room with an electric fan running is potentially lethal, even to the point that many Korean doctors and safety agencies formally warn against doing so. Scientists outside of Korea, however, easily dismiss the deaths as misdiagnoses of other conditions, and handily debunk the proposed mechanisms for the danger as implausible. But some Koreans have countered that there must be some explanation unique to Korea or Koreans: something to do with physiology, geography, or even their particular electric fans. Could Koreans be right that there is something more to this urban legend than mere tradition and confirmation bias?

Korean fan death isn't very old; not even going back as far as the use of electric fans in the country. The first electricity was installed at Gyeongbok Palace in 1887, just a few years after Schuyler Wheeler made the first two-bladed electric fans commercially available. By 1900, companies like Toshiba were manufacturing and selling electric fans throughout Asia. Despite nearly a century of history of usage without incident, in the 1970s the Korean media suddenly began reporting cases of fan death. They happened in the summer, in a closed room, and usually involved an elderly person sleeping alone, with an electric fan running in the room. In the morning, the victim would be found dead, with the only evident cause of death being the electric fan still sitting there, blowing its supposedly lethal breeze.

The situation today is that government safety agencies warn that fans must be used safely. The Korea Consumer Protection Board analyzed reports of heat-related injuries during the summer months for the three years prior to 2006, and made recommendations to address the five most often recurring dangers, with the first on the list being that doors should be left open when using electric fans or air conditioning.

> *If bodies are exposed to electric fans or air conditioners for too long, it causes bodies to lose water and hypothermia. If directly in contact with a fan, this could lead to death from increase of carbon dioxide saturation concentration and decrease of oxygen concentration. The risks are higher for the elderly and patients with respiratory problems.*

> *From 2003–2005, a total of 20 cases were reported through the CISS involving asphyxiations caused by leaving electric fans and air conditioners on while sleeping. To prevent asphyxiation, timers should be set, wind direction should be rotated and doors should be left open.*

Other government agencies give similar guidelines. As a result, many electric fans available in Korea today are equipped with an automatic timer feature, to make the fan turn itself off after a period of time for safety reasons. Fans from the Korean manufacturer Shinil Industrial bear a warning label that states "This product may cause suffocation or hypothermia". Such fans have a little warning sticker, a red circle with a yellow center, showing a body laying beside an electric fan.

The obvious questions to ask are how and why does this happen? But the important rule of thumb we learn to remember here is that the first question we should ask when investigating a strange phenomenon is does it actually happen at all? Obviously people do die in Korea, and some of them have fans running at the time. Whether the fan is causing the death is a question for the Korean coroners; and whether there's a plausible mechanism for such a death is a question for science. Let's take a quick look at the proposed explanations.

When we do this, we encounter our first red flag. Proposed explanations for Korean fan death are all over the map. There is no accepted scientific explanation. The most common is asphyxiation, caused by wind currents. Hypothermia is the second strongest contender, caused by the fan evaporating enough sweat off the victim's skin to wick away enough heat to kill them. But Korea's temperate climate is wet year round, especially during its summer monsoon season. The summers are hot and wet, and the winters are cold and comparatively dry. Fan death is a summer phenomenon; it's hot and wet. Are people really dying of hypothermia in such conditions? It is possible, if a stretch, if the fan's current is steady and dry enough. But fans don't dry the air or change its temperature; they merely circulate it.

Some posit really far-out explanations based on purely bad science. Some think the fan blades chop the oxygen molecules in half, rendering them useless or even poisonous; others believe that electric fans use up oxygen and produce carbon dioxide.

And, as with many urban legends, there's even a conspiracy theory that's been proposed to explain fears of fan death. When the stories first appeared in the 1970s, it was during a time when energy usage was sharply on the rise, and the Korean government thought up and spread the rumor to scare people into turning off their fans at night and saving energy. But like so many unproven conspiracy theories, this one has a pretty large hole in its logic. Energy usage during the summer is greatest during the day; and if the government *was* going to invent a rumor designed to reduce consumption, it would have made more sense to get people to turn something off during the day, like air conditioning or lights.

At least one Western expert has endorsed the theory that it is indeed the effect of the fan that kills. Dr. Laurence Kalkstein at the University of Miami is a climatologist and biometeorologist, studying the effects of weather on plants and animals including humans. At a conference in Korea, he explained that a fan blowing on an elderly person sleeping in a hot room would actually dehydrate the skin, causing death from respiratory distress. Even though the air blown onto the victim is as humid as a Korean monsoon, it should still carry away some amount of sweat from the skin, causing dehydration. So far as I've been able to find, Kalkstein has not examined any victims of fan death, so his suggested explanation remains unconfirmed.

However, there are doctors who *have* examined fan death victims. In 2007, Dr. John Linton who had autopsied several such people, told the International Herald Tribune:

> *There are several things that could be causing the fan deaths, things like pulmonary embolisms, cerebrovascular accidents or arrhythmia. There is little scientific evidence to support that a fan alone can kill you if you are using it in a sealed room. Although it is a common belief among Koreans, there are other explainable reasons for why these deaths are happening.*

Dr. Lee Yoon Song told the Korea Times in 2006:

> *Korean reporters are constantly writing inaccurate articles about death by fan, describing these deaths as being caused by the fan. That's why it seems that fan*

deaths only happen in Korea, when in reality these types of deaths are quite rare. They should have reported the victim's original defects such as heart or lung disease, which are the main cause of death in these cases.

But these dissenting opinions are in the minority. Most fan death victims examined by Korean doctors are given a cause of death of asphyxiation caused by the fan. Two professors of emergency medicine at Seoul's Samsung Medical Center agree that when a fan blows on your face, air currents develop that reduce the atmospheric pressure in front of your face by as much as 20%, causing a similar drop in oxygen availability. The victim then dies from the lack of oxygen. This is the prevailing view among Korean doctors who accept the fan death diagnosis.

Actual warning label

The view is also wildly implausible, from any number of basic science perspectives. First, we don't observe people keeling over dead on mildly breezy days, even if they're sitting on a beach facing into the breeze for a long time. Second, we have no cases of brain damage from asphyxiation that was not quite sufficient to kill, which should be far more prevalent if this were indeed happening. Third, wind striking you in the face does not *reduce* the pressure at the front of your head; it *increases* it. Fourth, 20% is a huge pressure differential. 15 kph of wind would create around 55 grams of pressure on the average head, which is less than 1% of 1% more than normal atmospheric pressure. The Samsung doctors' 20% drop in pressure on one side of your head would require in the neighborhood of 100 kilograms of force at sea level, which would require a wind speed of at least 650 kph. So if you're using an F-18 fighter jet engine for a fan, it starts to look plausible.

So how do we properly analyze the phenomenon of Korean Fan Death? If we follow a truly skeptical process, what do we come up with? We've looked at the data and found that there are, indeed, plenty of people who have been found dead near a fan and had their deaths classified as fan death. But we also have good reasons to suspect that those causes of death were misdiagnosed: there are simply no plausible mechanisms for the breeze from a fan to be lethally dangerous. There was no significant difference in fan design introduced around the time the deaths began to be reported, and no differences between Korean fans and those in the rest of the world. There is nothing unique to Korea's geography (that we know of) that would explain why such a thing happens only there, and there aren't really any comparable cases of unique geography elsewhere in the world making certain technologies dangerous. Koreans outside of Korea don't seem to have any trouble with fans, and there is no known difference in Korean anatomy that would make them especially susceptible. Truly, all the possible factors that would make fan death a uniquely Korean phenomenon fall apart under scrutiny.

However, there's at least one remaining possibility that can explain what's being reported, and it doesn't require any new discoveries about anatomy or fans, or any special conditions. The simple fact is that we absolutely expect to see a correlation between summer deaths and fan usage. When it's hot and muggy in the summer, people are going to be running their fans; and when high-risk elderly people happen to die from whatever heat-related cause, it's perfectly likely that a running fan will be found nearby. The perception of a causal relationship between the two will be reinforced every time it's confirmed by another such body being found. This simple confusion between correlation and causation adequately explains all twenty diagnoses investigated by the Korea Consumer Protection Board, and it explains the convictions of the doctors, the fan manufacturers, and the safety boards.

So we never really get past our original question: Whether fan death has ever actually happened. It may have, most likely via hypothermia, but it would have to be in a colder, drier climate, and would be evenly distributed throughout such regions of the world. But so far as the existence of a specific Korean fan death phenomenon, the skeptical mind concludes no, there is no good evidence for such a thing's existence.

REFERENCES & FURTHER READING

Editors. "Why Do Koreans Think Electric Fans Will Kill Them?" *Esquire.* Hearst Communications, Inc., 22 Jan. 2009. Web. 1 Dec. 2011. <http://www.esquire.com/style/answer-fella/korean-fan-death-0209>

Editors. "The Truth of Fan Death?" *Associated Press.* 16 Jul. 2008, Newspaper.

Mikkelson, B. "Fan Death." *Snopes.com.* Barbara and David P. Mikkelson, 6 Jul. 2011. Web. 3 Dec. 2011. <http://www.snopes.com/medical/freakish/fandeath.asp>

Office of Public Relations. *Beware of Summer Hazards!* Seoul: Korea Consumer Protection Board, 2006.

Shin-who, K. "Do Electric Fans Cause Death?" *Korea Times.* 10 Sep. 2006, Newspaper.

Surridge, G. "Newspapers fan belief in urban myth." *International Herald Tribune.* 10 Jan. 2007, Newspaper.

32. PIT BULL ATTACK!

*Pit bulls have a reputation for being the most danger-
ous dog breed. Is this reputation deserved?*

Perhaps the most horrifying story to hear on the news is a
case of a child being killed by a pack of dogs, hardly anything
can incite a more emotional response. We're quick to vilify the
dogs; perhaps justifiably so, perhaps not. In the United States,
it seems that more often than not, the dogs involved in such
attacks are pit bulls. Legislation is quick to address highly emotional issues, and many states now have various bans and limitations on pit bulls. Today we're going to turn our skeptical eye
onto the popular belief that pit bulls are truly as dangerous as
their reputation suggests.

Defining a pit bull is not exactly a slam-dunk. It's not a
specific breed; rather it's a collection of several related breeds.
Those dogs that are unambiguously pit bulls include the American pit bull terrier, the American Staffordshire terrier, and the
Staffordshire bull terrier; however the bull terrier or English
bull terrier is not. Some municipalities classify the American
bulldog as a pit bull "type" dog. Finally, pit bulls are usually
classified as any dog having the substantial physical characteristics and appearance of pit bull breeds, which establishes the
somewhat unfortunately vague precedent that you know a pit
bull when you see one.

Deaths by dog attack have been thoroughly studied. Perhaps the most often cited large study was published in 2000 in
the *Journal of the American Veterinary Medicine Association* assessing 20 years of DBRF (dog bite related fatalities) in the
United States, from 1979 through 1998. During that period,

238 Americans were killed by 403 dogs. Just over half of these deaths involved pit bull type dogs and Rottweilers. It's important to note that there is always some uncertainty about breed. A lot of dogs out there are not pure bred or are mixed,

and numbers for those dogs were included in the study as well. But the trends over 20 years were clear: Pit bulls are indeed responsible for the most DBRFs, though in some years Rottweilers were most deadly. German shepherds are third, huskies and malamutes are next, and it goes down from there. Pit bulls killed more than seven times as many people as Doberman pinschers, which we usually consider to be so dangerous.

The authors of the study also noted one very important weakness of such studies: they look only at the dogs themselves, and not at the owners. The example they give is that of an owner who wants an aggressive dog, perhaps as a guard dog, or as an ornament for his barbed-wire bicep tattoo. An owner who wants a scary dog, and who plans to use it in a macho or antagonistic way, is much more likely to buy a pit bull to put into this role than he is a poodle or Chihuahua. Some percentage of potential dog bite scenarios are always going to be set up by aggressive dog owners; so statistically, we're always going to see a correlation between dog bites and certain breeds that were selected based on reputation, whether that reputation is deserved by the dog or not.

When lacrosse coach Diane Whipple was killed by two pit bull type dogs in San Francisco in 2001, the specific breed was a Presa Canario. Sales of these shot up, driven by people who wanted the latest and greatest bad-boy dog. They were selected by aggressive people based on reputation. Indeed, the San Francisco dogs were owned by a couple who was raising them on behalf of prison inmates trying to run a dog fighting operation from their prison cell.

The other glaring weakness of this type of study is that it doesn't take into account the relative prevalence of each of these dog breeds. Maybe pit bulls killed seven times as many people as Dobermans because there are seven times as many of them out there. So what are the numbers; are pit bulls, Rottweilers, and German shepherds sufficiently popular that numbers alone can account for the number of deaths they cause?

This is, unfortunately, a question that cannot be well answered. The only real manifest of dog breed popularity in the United States is the American Kennel Club's registry. This registry includes only dogs that owners choose to register, and is highly skewed toward pure bred dogs owned by serious dog owners. It does not include anywhere near the more than 75 million dogs living in the United States. Labrador retrievers nearly always top the AKC list, yet these were listed 12th in fatalities in the 2000 study. Rottweilers, found right at the top of the DBRF list, rank down in the teens on the AKC registrations. The relative registrations of German shepherds, on the other hand, does match their position on the DBRF list, right up in the top two or three.

We might be able to make the following conclusions from these few data points:

1. Labrador retrievers are very safe dogs.

2. Rottweilers are very dangerous dogs.

3. German shepherds are about average.

But whether these conclusions are true or not, we don't have enough data to confirm them, because the AKC registration data does not necessarily reflect the actual prevalence of these dog breeds out in the world.

Neither should fatalities be considered a significant factor when assessing how dangerous a certain dog breed is. Another highly cited study was published in the journal *Pediatrics* in 1994, and it found an average of about 20 deaths a year from dogs in the United States, compared to *585,000* dog bite injuries requiring medical treatment. Dogs inflict injuries nearly *30,000 times as often* as they inflict death. Clearly, injury is where the overwhelming majority of dog aggression is, not death; and it's probably where we really should be looking to determine the relative aggression of certain breeds.

The authors surveyed nearly 1,000 reports of dog bites in the city of Denver in 1991, and restricted their results to cases where they were able to contact the owners and get complete information about the dog, its history, and the circumstances of the bite. Then, for each biting dog, they found a geographically nearby control dog, of any breed, with no biting history. Dog breeds were reported by the owners, and in cases of mixed breed, dogs were listed as whatever breed the owner considered to be dominant. Since the non-biting control dogs were a random selection from the existing breed distribution in the same region as the biting dogs, the factor of breed prevalence was effectively canceled out. These authors structured their study to give us a real picture of which breeds, or other factors, most often contribute to dog bites. And here's what they found.

Surprise: German shepherds and chow chows are the big biters. Golden retrievers and standard poodles are the least likely to bite. Dogs whose distribution among the biting and non-biting populations was not significant include Chihuahuas, cocker spaniels, Dobermans, Labrador retrievers, Scottish terriers, and Shetland sheepdogs. For all other dog breeds, there was insufficient data.

But where are pit bulls in that list? When the study was done, new pit bull ownerships had been banned in Denver since 1989, so there were no pit bull bites recorded in the study. This ban was based on 20 pit bull attacks in Colorado over the preceding five years. That's four a year, out of a nationwide 585,000 a year. A class of plaintiffs called the Denver Dog Fanciers tried to overturn the ban, unsuccessfully. The court's findings included:

> *It cannot be proven that pit bull dogs bite more than other dogs. However, there is credible evidence that pit bull dog attacks are more severe and more likely to result in fatalities.*

And:

> *The City did prove that [pit bulls] inflict more serious wounds than other breeds. They tend to attack the deep muscles, to hold on, to shake, and to cause ripping of tissues. Pit bull attacks were compared to shark attacks.*

These points do seem to be supported by the facts. Pit bulls are involved in a disproportionately high number of fatal versus non-fatal attacks, though this number is still extremely small. Pit bulls do tend to bite and hold, displaying an amazing ability to not release their grip. This has given rise to the rumor — which is completely false — that they have some physiological ability to "lock" their jaw. There's also no truth to the story that pit bulls have uniquely large jaw muscles, or have the highest measured bite strength. Pit bulls are strong, no doubt about it; but so are many other large dogs.

Only a very few studies of dog bite force have been done, and Rottweilers seem to be the strongest found so far, with a bite force of around 1,400 newtons. Pit bulls have been measured at 1,100 newtons. For comparison, hyenas can bite with four times the force, at over 4,400 newtons. But keep in mind that these numbers are from very small pilot studies.

Here are some other factors that the *Pediatrics* authors found. Dogs bite more often when they're male, when they're not neutered, when they're over 20 kilograms, and when they're less than five years old. Biting dogs are more likely to live in homes with children below the age of ten, are more likely to be kept chained when outdoors, and are more likely to growl at visitors. Interestingly, obedience training, guard training, and discipline styles have not been found to have a statistically significant impact on that dog's likelihood to bite.

So here's the bottom line, based on my own analysis of the available data. If you want a safe dog, avoid chow chows and German shepherds. Golden retrievers are your best bet. Pit bulls may well be a breed to avoid, but there is not definitive data to support this. Get a female or a neutered male, small, and over five years old. The fewer children around, the less likely it is to bite.

If a dog *is* going to bite you though, the two breeds you least want it to be are a pit bull or a Rottweiler. They are definitely the most dangerous biters, once they decide they're going to bite you. If you see one on the street, there is not sufficient data to support any particular need for concern. Like all dogs, its owner and its environment are major factors in its level of aggression.

This is a case where the value of good science is to drive policy. Most researchers agree that breed-specific legislation — a nice term for pit bull bans — are inappropriate. No good data exists to demonstrate that such bans have had any impact. Improved enforcement of existing laws, and improved education for dog owners, are far more likely to reduce the number of dog bites, fatal or not.

REFERENCES & FURTHER READING

Gershman, K., Sacks, J., Wright, J. "Which Dogs Bite? A Case-Control Study of Risk Factors." *Pediatrics*. 1 Jun. 1994, Volume 93, Number 6: 913-917.

HSUS. "Dangerous Dogs and Breed-Specific Legislation." *The Humane Society of the United States*. The Humane Society of the United States, 10 Feb. 2010. Web. 10 Dec. 2011. <http://www.humanesociety.org/animals/dogs/facts/statement_dange rous_dogs_breed_specific_legislation.html>

Lindner, D., Marretta, S., Pijanowski, G., Johnson, A., Smith, C. "Measurement of bite force in dogs: a pilot study." *Journal of Veterinary Dentistry*. 1 Jun. 1995, Volume 12, Number 2: 49-52.

Nelson, K. *Denver's Pit Bull Ordinance: A Review of Its History and Judicial Rulings*. Denver: Denver City Attorney's Office, 2005.

Sacks, J., Sinclair, L., Gilchrist, J., Golab, G., Lockwood, R. "Breeds of dogs involved in fatal human attacks in the United States between 1979 and 1998." *Journal of the American Veterinary Medical Association*. 15 Sep. 2000, Volume 217, Number 6: 836-840.

Swift, E. "The Pit Bull: Friend and Killer." *Sports Illustrated*. 27 Jul. 1987, Volume 67, Number 4.

33. THE MYSTERY OF THE MARY CELESTE

The facts, as we know them, about what really happened to maritime lore's most famous missing crew.

In 1872, a ship was found adrift in the Atlantic Ocean, in near-perfect condition but for one problem: there was nobody aboard. In time, the story of the *Mary Celeste* became one of the most famous riddles of the sea. Over the years, many have offered solutions for what happened to the crew. But are any of them correct?

As is the case with so many of the mysteries we examine here on *Skeptoid*, the story of the *Mary Celeste* was an actual event that was largely forgotten until an imaginative author revived and exaggerated it for popular audiences. This time, the author was a young man who would later be knighted as Sir Arthur Conan Doyle for his *Sherlock Holmes* books. It was a short story written under the pseudonym W. Small for the January 1884 issue of *Cornhill Magazine,* entitled "J. Habakuk Jephson's Statement". Conan Doyle dramatized the *Mary Celeste's* story, adding such touches as meals laid out on the table, tea boiling on the stove, and the ship sailing boldly into the harbor at Gibraltar with nobody at the helm. Today, most people who have heard of the ship think these details are part of what actually happened. They aren't.

Conan Doyle's was only the first of many such treatments. A 1913 magazine article was the forged account of a man named Fosdyk who claimed to have been a stowaway on board the *Mary Celeste,* witnessed the entire crew fall overboard as they pressed against the rail to watch three of the men have a swimming race, then managed to be the only one not eaten by sharks and eventually washed ashore on Africa. In the 1920s, an author named Keating forged an article for *Chamber's Journal* telling the story of a man named Pemberton who survived. Keating soon expanded the fictitious Pemberton's tale into a book called *The Great Mary Celeste Hoax*. Unfortunately, the book's success became its downfall: Interviews with Pemberton were widely sought. Keating tried to weasel his way out with excuses, and even offered a picture of his own father as a photograph of Pemberton; but it was soon discovered that he made the whole thing up.

The *Mary Celeste* was a small merchant brigantine of 33 meters and 282 gross tons. She'd just been acquired by a small group of investors, among whom was the ship's one-third owner, Captain Benjamin Briggs. Joining him on board were his wife and baby daughter, plus seven sailors. They left port from Staten Island, New York in November of 1872, fully laden with cargo bound for Genoa, Italy. The cargo was 1,701 wooden barrels of pure grain alcohol, intended to fortify cheap Italian wines. America's vast corn fields made it the cheapest producer of grain alcohol at the time, and it made good economic sense for Italy to buy it and ship it all the way from the United States.

The voyage was relatively uneventful according to Captain Briggs' log entries, and the fine weather was confirmed by the captain of another ship sailing one week behind. Captain David Morehouse commanded the *Dei Gratia,* a similar brigantine laden with petroleum. Briggs and Morehouse had sailed together for many years and knew each other well, and it was a happy coincidence that the two friends found themselves on nearly identical voyages.

But almost halfway between the Azore Islands and Gibraltar, Morehouse made an unhappy discovery. The *Dei Gratia* unexpectedly caught up with the *Mary Celeste,* finding her adrift. Morehouse sent a party to investigate, and found the *Mary Celeste* unmanned. It was a bizarre find; there were no obvious signs of trouble and all appeared to be in order. But there were a few interesting clues.

The *Mary Celeste* had been equipped with a yawl, and though that term usually describes a type of sailboat, in this case it refers to a ship's rowboat capable of being rigged for sailing. The yawl was normally stored atop the main cargo hatch between the two masts, but was gone; and the railings on one side of the ship had been lowered indicating that the yawl had been launched normally. The other two cargo hatches — the forehatch on the foredeck and the lazarette hatch, above a small compartment aft — had both been removed and were stowed, exposing the cargo of alcohol.

When Morehouse found the ship, it was flying minimal sails, the fore lower topsail and two jibs. Modern analysis has confirmed that Morehouse found it just about where it would have been expected to be, driven primarily by currents, if it had been under no helm control since passing the Azores. Interestingly, the main peak halyard, the stoutest line on the ship, was missing; and it was very likely the same rope that was found cleated off and trailing in the water behind the ship. There was a significant amount of water in the bilge and cabins of the ship, but this was believed to be consistent with the open hatches and an opened skylight. The *Mary Celeste* had tossed about for at least ten days since its last log entries, in freshening weather that had compelled the crew to shorten sail; and so it was not surprising that it had taken on some water. As its stores were in good shape and it was perfectly seaworthy, Morehouse sent a skeleton crew aboard the *Mary Celeste* and brought it to Gibraltar, where the loss was reported and investigations took place for purposes of insurance and salvage. When the cargo was unloaded and examined, nine of the barrels of alcohol were empty: undamaged, yet empty.

Early theories quickly focused on the relationship between Briggs and Morehouse, and charges of conspiracy and insurance fraud were flung about; but these theories made no sense from a profit standpoint. For a while, some believed piracy had taken place, or that perhaps Briggs' crew had drunkenly mutinied against him; but all of these stories crumbled under scrutiny and lack of evidence that would have been expected.

Since then, even more suggestions have come from the fringe, pointing to exotic causes for the abandonment, like waterspouts and rogue waves. One in particular, David Williams, proposes that a "seaquake" struck the ocean floor. He states that the US and British navies know that such quakes can destroy surface vessels with powerful shockwaves, but that they cover it up so it's not generally known. Williams' theory is that this sudden shaking released embers from the ship's stove, so the crew fled the ship fearing the embers would ignite the store of alcohol. Williams states that there are numerous examples of ships sustaining heavy damage from such shockwaves, but as no damage was noted on the *Mary Celeste*, his seems an arbitrary explanation.

A few pieces of physical evidence strongly suggest what has emerged as the favorite theory, based on those nine empty barrels discovered in Gibraltar. The reason they were empty would have been clear to any cooper. All of those many barrels were of white oak, except for those nine, which were of red oak. Of the species of wood sold as white oak, the majority have occluded pores. This makes the wood watertight, which is why white oak is used for wine barrels and other barrels intended to hold liquid. The pores in the wood of the twenty or so species of red oak, on the other hand, are open; allowing liquids to seep through the wood. Consequently, red oak barrels should only be used for dry goods. But, for some reason, Meissner Ackermann & Co. (owner of the alcohol) used nine of the wrong type of barrel.

At some point in the voyage, or possibly even before, these barrels would have become soaked through. Alcohol evaporates quite quickly, so the smell would have permeated the ship's cargo hold. No record remains of where in that vast pile of 1,701 barrels the nine red oaks were found, but chances are that most of them were hidden from view. A visual inspection of the cargo hold probably would have found nothing, making it impossible to tell the extent of the leakage, but that smell would have been everywhere. This is also evidenced by the removed deck hatches; the crew were undoubtedly trying to vent the flammable fumes. But alcohol vapor is heavier than air, so it's unlikely that venting the deck hatches would have done much to dispel it.

A red oak barrel of the type aboard *Mary Celeste*

The crew had to have feared that an explosion or fire was imminent. The yawl was launched and everyone on board removed to it. This was not done in a panic or haphazardly, but rather urgently and efficiently. The captain had the sense to collect his sextant and marine chronometer, necessary for navigating; but everything else on board that was not essential was left behind. No strong evidence suggests an answer to the question of whether they intended to completely abandon the ship, or to simply sit at a safe distance until they figured the danger was past. They took the precaution of using the strongest line they had to secure the yawl to the *Mary Celeste*, but in some unknown circumstance, the line was not secure or became severed. The few sails still set on the ship were enough that the yawl's rowers could not keep up. Once he saw they would not be able to catch the ship, the captain headed for Santa Maria Island in the Azores. And, as was all too often the end of such deep sea open boat voyages in those days, they never made it, and were ultimately swallowed by the Atlantic Ocean.

Over the next ten days, the *Mary Celeste* rocked in the breeze with its open hatches. The last of the alcohol evaporated away, and no one in Morehouse's party reported smelling anything. It's a certainty that all nine barrels, some 450 Imperial gallons, escaped as fumes while the *Mary Celeste* was at sea.

In 2006, Dr. Andrea Sella, a chemist at University College London, conducted an experiment to recreate conditions that he believes may have prompted Briggs to evacuate. Sella filled a compartment with cubes of paper and butane gas, then sparked it. The resulting combustion produced a sudden flash of flame that was visually dramatic, but was cool and quick enough that the paper was not scorched. Dr. Sella theorized that perhaps such a flash had happened in the *Mary Celeste's* cargo compartment, frightening the crew into fearing that a much larger explosion may well have been imminent. The ethanol vapors in the *Mary Celeste's* hold would burn even cooler and quicker than butane, though probably much less dramatically, with a blue or invisible flame, unlike like the butane's yellow flash. But it certainly would have been every bit as alarming to the crew, if it had happened.

Without any reasonable doubt, the cause of the disappearance of the *Mary Celeste's* crew was voluntary abandonment. We can't be certain what prompted the evacuation, but there seems little reason to speculate beyond what's best supported by the evidence: powerful and dangerous fumes from the alcohol-soaked red oak barrels. Briggs' action, though ultimately disastrous, was more than reasonable at the time.

References & Further Reading

Begg, Paul. "The Classic Case of the Mary Celeste." *The Unexplained Mysteries of Time and Space.* 1 Jan. 1982, Volume 4, Issue 48.

Blumberg, J. "Abandoned Ship: the Mary Celeste." *Smithsonian.* 1 Nov. 2007, Volume 38, Number 8.

Corrado, J. "What Really Happened to the Mary Celeste?" *The Straight Dope.* Creative Loafing Media, Inc., 16 Oct. 2001. Web. 17 Dec. 2011. <http://www.straightdope.com/columns/read/1962/what-really-happened-to-the-em-mary-celeste-em>

Doyle, A. "J. Habakuk Jephson's Statement." *Cornhill Magazine.* 31 Dec. 1884, Volume 2, Number 7: 1-32.

Lee, A. "Solved: The Mystery of the Mary Celeste." *UCL News.* University College London, 20 May 2006. Web. 17 Dec. 2011. <http://www.ucl.ac.uk/news/news-articles/inthenews/itn060522>

Wengert, G. "Red Oak, White Oak, Black Oak, and More." *WoodWeb.* WOODWEB Inc., 20 Jun. 2005. Web. 17 Dec. 2011. <http://www.woodweb.com/knowledge_base/Red_Oak_White_Oak_Black_Oak_and_More.html>

34. APPROACHING A SUBJECT SKEPTICALLY

My process for examining a new topic, to learn whether it's fact or fiction.

One of the questions I get asked a lot is how I go about approaching a new subject. When you hear about something new, what's the best way to think about it? What's the best way to determine whether it's science or pseudoscience? Well, I'm not sure that there is a "best" way, and I don't think there's one methodology that everyone can follow that's going to work in every circumstance, but I'll try to give the best answer I can. It's probably not the same answer you'd hear from others, but this is what works for me.

First of all, and perhaps most important, is that there's a separation between my daily life and working on *Skeptoid*. I don't walk around demanding peer-reviewed scientific evidence for everything that I see. I don't have a crazed, obsessive drive to know the validity of every new product for sale at the mall. I'd never get through my day without a certain amount of tolerance for pseudoscience. Fad products, marketing campaigns, greenwashing, and even straight-up fraudulent claims surround us, all day, every day. I accept that. Trying to be a full-time challenger of pseudoscience would not only be hopelessly quixotic, it would also annoy everyone around me, and rob me of the freedom to enjoy my day.

So I let virtually everything slide. A coworker is wearing a magic bracelet? Great, good for him. Neighbor talks about her great visit to the reflexologist? Bully for her. Overhear some people discussing what Nostradamus said about the 2012 apocalypse? Whatever floats their boat.

But what if I'm out with friends and somebody asks me my thoughts on something? This happens all the time. You're on the spot, you don't have access to research materials, you don't have time to look into it. Now, oftentimes I've already written about the subject in question, or something really similar, that gives me a pretty good foundation. Sometimes I haven't, and like most people, have to rely on a journeyman's knowledge of a subject area that's outside of my core competence. This provides a pretty good overview of whether or not the new claim is in line with what's generally known about the subject. Usually it's not; otherwise it wouldn't be on the news or wherever it was that my friends heard about it.

So there you are. You're given something that raises your skeptical radar, it's outside your core competence, and your friends just saw it on television or the Internet. Despite the fact that most people *say* they take TV or Internet reports with a grain of salt, few actually do. There's something deeply compelling about hearing a claim from an authoritative source; we all have a voice in the back of our heads that wants the new claim to be true, and this desire gets confirmed by the belief that the story wouldn't have made it all the way to the TV news without having been pretty well substantiated. What are you going to do?

The first thing I'd do is take out my phone and track down the original source of the story, using keywords from the report to search Google. I'd want to know if it was reported in any journals, or if it skipped this process and went straight to the mass media. This is the simplest and fastest way to see if a new claim or phenomenon has come from the world of legitimate research, or if it comes from a crank, charlatan, or manufacturer operating outside of science. You always have to remember that the mass media doesn't care; they're interested in the sensationalism of the story, not in its validity.

That's it. That's probably all I'm going to do when I'm out in the world and get a question that's worthy of looking into. It's not a perfect process, but nine times out of ten this will correctly tell you whether there's something there, or whether it's just more noise from media clamoring for eyeball share.

It's only when I take my seat in the *Skeptoid* office that I assume the mantle of proper separator of fact and fiction. This is when I take each week's topic and give it my honest best effort at a good skeptical treatment. The best topics are those that are popularly misunderstood, but with facts behind them that, when properly understood, are way cooler than the popular version. This isn't as hard as it might sound; nearly every popular myth has some history that puts its genesis into a fascinating new perspective.

Sometimes finding this perspective takes me back in time, to an out-of-print book, or to a newspaper article a century old. Tracking these down requires a lot of eBook purchases, Google Books downloads, newspaper archive searches, and occasionally even the coveted trip to a real library to find a real book. Of course, even the relevant pages from the real book end up as electronic files on my computer, photographed with the iPhone and then OCR converted to searchable text. Getting brand new information, like current research, is almost exactly the same process; it's all available when you have the right accounts to access online research libraries. But none of that compares to the few chances to actually go in person to a place where something strange is said to have happened: to smell the dusty desert wind across Death Valley's Racetrack Playa, to touch the cold granite of the Georgia Guidestones, and to photograph a Fata Morgana mirage such as the ones responsible for so many legendary ghost lights.

I've been doing this show every week for five years now, and on the one hand, you might assume that I've developed a certain aptitude for smelling rats, and have pretty good radar for science vs. pseudoscience. That's true to a degree; but at the same time, I've learned that I can easily be surprised. I often learn that something that sounded pretty hokey is actually true, and something I took for granted turns out to be false. So rather than having developed a supersense for fact and fiction, I've actually picked up a more acute awareness of my own ignorance. Kind of the opposite of what one might hope for; but as we see so often, magically easy solutions to complex problems are a fool's gold.

The process is different every time, but it always starts with a quick survey of the most popular sources, followed by delving deeper into the roots. If it's homeopathy, I want to know what led Samuel Hahnemann to his original conclusions. If it's a conspiracy theory, I want to know who came up with it and what question they were trying to answer. If it's a ghost story, I want to know who first wrote about it and what their relationship was to the hauntee. It's critical to allow for the possibility that the story may or may not be as reported; and to follow up the leads in both directions. Frequently this requires some pretty detailed departure from the popularly known core of the story.

For example, say you find a reference to the mayor of an old town. First you find out if the town actually exists, where it was, whether it's still there, and find it on Google Earth to see if it makes sense within the context of the story. Then find out if the person listed as the mayor actually was the mayor. Find out when he was born, see if the timing is right. There are myriad details you can drill down through, to be as thorough as possible validating the story. Sometimes there are an endless number of these leads, and with only a week between podcast episodes, I often have to simply stop following them, thus making many episodes necessarily incomplete and open to error.

But when you have the time, how far do you go tracking these leads? I've found that there's never a point of diminishing returns. Every time I've made a discovery or connection that (to my knowledge) no other researcher has found, it's always in one of these fine tails of data. The unturned stones are rarely in the middle of the road most traveled. They're in the obscure newspaper article that never got syndicated; they're in the out-of-print interview with the expert who was misquoted in the popular version of the story; and more than anywhere else, they're in the actual published research that was omitted from the mass media reports because it did not support a sensational revisioning of the story.

I don't mean to sound cynical about the mass media. There are many, many excellent reporters and news bureaus who conscientiously produce exceptional material. But I think you'll find that the better they are, the more likely they are to give you an honest assessment of the industry's overall goal, which is to be profitable. The easiest way to do this, as practiced by a probable majority of editors, is to be sensational. I don't think it's a cynical assessment, and it has certainly proven itself to me time and time again through my work validating mass media reports.

So take the road most traveled, as presented in Wikipedia, to get the lay of the land. But to truly learn anything new, you must explore those obscure details that nobody else had time for, or that they overlooked.

Interestingly, I'd say that my process — though it's much more thorough — is probably no more accurate than the quick trick in the restaurant with a smartphone and Google. The more information I collect, the more possibility for error. The more obscure threads I follow, the more are likely to be unreliable. And the more time I spend trying to be thorough on one part of the story, the less time I have for the other parts: an unfortunate exigency of producing a weekly show. I'd say that errors of omission are my most common mistakes, followed by errors that I just didn't catch because of limited time. And like every fallible biological entity, I also make errors by misinterpreting, misreading, and failing to see beyond my own personal biases.

You'll make these same errors in your own research. The best defense against them is to acknowledge your blind spots, compensate for them, and honestly qualify remarks that you can't be sure of. First I try to be right more often than I'm wrong, but second I try to emphasize the process over the conclusions. Being right nine times doesn't guarantee that you'll be right the tenth time, but trying hard all ten times guarantees that you'll at least be as right as your process is capable of.

REFERENCES & FURTHER READING

Nickell, J. *Real or Fake: Studies in Authentication.* Lexington: University Press of Kentucky, 2009.

Plait, P. *Bad Astronomy: Misconceptions and Misuses Revealed, from Astrology to the Moon Landing Hoax.* New York: Wiley, 2002.

Radford, B. *Media Mythmakers: How Journalists, Activists, and Advertisers Mislead Us.* Amherst: Prometheus Books, 2003.

Randi, J. *Flim-Flam! Psychics, ESP, Unicorns, and Other Delusions.* Buffalo: Prometheus Books, 1982.

Sagan, C. *The Demon-Haunted World: Science as a Candle in the Dark.* New York: Random House, 1995.

Shermer, M. *Science Friction: Where the Known Meets the Unknown.* New York: Times Books, 2005.

35. THE TOXIC LADY

In 1994, fumes from a woman's body knocked out most of an emergency room staff. What happened?

Television news lit up in the United States in February of 1994 when a 31-year-old woman, Gloria Ramirez, died in a hospital emergency room. She'd been acutely ill with advanced cervical cancer, and when she began having pulmonary and respiratory problems, she called paramedics. Soon after she was brought to the emergency room at Riverside General Hospital in southern California, she passed out, and never regained consciousness. So far, there had been nothing unusual or medically out of the ordinary.

One the nurses drew blood from Ramirez, and noted that it both looked and smelled strange. It had an ammonia-like odor, and several people noticed manila-colored crystals floating in the blood. While the emergency room fought to reverse Ramirez' rapidly deteriorating condition, some of the staff began falling ill. Symptoms included dizziness and fainting, a sensation of burning on the skin, nausea, apnea, tremors, even paralysis. Ramirez died, and as her body was moved into isolation, those attending her also fell ill. The emergency room was evacuated to the parking lot. In all, 23 people became ill. Five were hospitalized. One nurse was kept in the hospital for ten days with tremors and apnea. The most seriously ill, a doctor in residence named Julie Gorchynski, stayed in intensive care for two weeks, contracting apnea, hepatitis, pancreatitis, and necrosis of the bone marrow which crippled her legs for months and required at least three surgeries.

Television crews arrived about the same time as the Riverside County hazardous materials team, and as southern California flew into a panic that a woman's body was knocking people out with its fumes, it was hardly noticed that the hazmat team came up empty handed. They found nothing unusual inside the emergency room. They searched for every kind of toxic substance they were equipped to find, and detected nothing that could account for the staff illnesses.

The task fell to the coroner, whose pathologists were charged with autopsying the toxic body. It was the most unusual autopsy the county had ever seen: doctors wearing full airtight suits with respirators, in a special sealed room. They took samples of everything: her tissue, her blood, even air from the bodybag she'd been in. And the final analysis? Nothing. The coroner's office found nothing inconsistent with a victim of cervical cancer, and like the hazmat team, nothing that would have knocked out the hospital staff or been harmful in any way.

The toxic lady, it turned out, was not toxic at all, by the all measures the doctors knew to employ.

And yet two members of the emergency room staff still lay in the hospital with undeniable physical medical conditions, and the rest of the staff all recalled the odors and strange looking blood. Something real had happened that night in February, and all the signs were that Gloria Ramirez, or something inside her, was the cause. Nevertheless, it couldn't be found with any certainty. Many investigations led to dead ends. And in September 1994, nearly seven months after the toxic lady felled the medical staff, the health department released its official report. Ramirez died from cervical cancer, and nothing else. The emergency room victims were found to be free of any explicable medical causes, and were determined to have suffered from a mass sociogenic illness, triggered by a frightening odor of unknown origin.

A sociogenic illness is one that is caused or influenced by social factors, rather than by a physical disease agent. It's a form of mass hysteria where the effect is a perceived illness. The concept of sociogenic illnesses is controversial, and labeling any event to be one always causes dissent and challenge. It's a diagnosis that almost nobody will accept.

During the 1990 gulf war when the first Iraqi SCUD missile struck Israel, 40% of the nearby civilians reported symptoms consistent with a gas attack, exactly as they expected; despite no chemical warhead being in the missile. In 1998, 800 Jordanian schoolchildren were vaccinated, and 122 were admitted to the hospital for what they believed were side effects; but for nearly all of them, no ill effects were found at all. Hundreds more schoolchildren fell ill in Belgium in 1999 after having drunk Coca-Cola, though nothing was found wrong with the beverage and none of the children had anything show up on blood tests. All three of these events are believed to be examples of sociogenic illness. In all cases, the epidemic was probably triggered by a very few victims who responded to some unknown triggering cause, probably a real reaction to something. But the cause was misinterpreted as whatever was obvious at the moment, and others who had been exposed to the same misinterpreted trigger experienced acute stress and fear, and the mass sociogenic illness was initiated.

Much about the Riverside toxic lady episode is consistent with this diagnosis. There were a number of startling surprises when Gloria Ramirez arrived at the emergency room. Besides the ammonia-like smell to her blood and the strange manila crystals in it, staff noticed that the skin on her abdomen had a weird oily appearance and smelled like garlic. Any one of these, or especially all of them together, might have triggered fear, nausea, or other stress responses in the nurse who drew the blood, which she and others may have interpreted as a physical response to toxic gas. And notably, almost all of the victims were female, and females are historically far more susceptible to sociogenic illness, according to a literature survey published in the *British Journal of Psychiatry* in 2002. No toxic residue was found by the hazmat team or by the coroner's team. The paramedics who had answered Ramirez' call and brought her to the hospital also came into contact with her blood when they started an intravenous line, and reported no ill effects at all. Despite its seeming improbability for an experienced emergency room staff, the sociogenic illness explanation was not only a good fit for the toxic lady incident, it was almost an open-and-shut case.

But in spite of the official report, toxicology investigations had been going on behind the scenes the whole year. When the coroners found nothing, they enlisted some outside help from an impressive source: the Lawrence Livermore National Laboratory in northern California. Since there wasn't much doing in the cold war business in 1994, Livermore had set up a forensics lab to offer their expertise to law enforcement agencies that might need it.

While on center stage the sociogenic illness explanation was offered, and Dr. Gorchynski filed a $6 million lawsuit against the hospital, backstage the Livermore team was hard at work. To make a long story short, they finally pieced together a scenario that some see as plausible, and some not so much. Their breakthrough came from gas chromatograph mass spectrometer analysis of the samples from Ramirez, and also from the headspace, which is the air between the sample and the lid of the container. The spectrometer showed one surprising peak that couldn't be accounted for by the drugs Ramirez had been given: a concentration of dimethyl sulfone.

Dimethyl sulfone is one oxygen atom away from a similar chemical, dimethyl sulfoxide, commonly called DMSO. DMSO is sold as a gel in hardware stores as a powerful degreaser, and it's also used by athletes to rub onto sore muscles. In fact, many people put it on their skin to relieve pain from conditions like arthritis. It's not really great for you, but people do it anyway. DMSO also caught the attention of the Livermore researchers because it would explain the greasy appearance of Ramirez' torso and the garlic-like odor.

When the paramedics gave Ramirez oxygen in the ambulance, the high oxygen concentration in her blood would have combined with the DMSO they theorize she had self-administered to relieve pain from her cancer, and formed the dimethyl sulfone observed in the spectrometer results. Ramirez' family insisted that she did not take DMSO, but the spike on the spectrometer is pretty hard to argue with, and she certainly would not have been the first cancer patient to do that. Moreover, her cervical cancer had caused kidney failure (which is actually what killed her), and any DMSO would have built up to very high levels in her blood. In the Livermore researchers' tests to reproduce the process, dimethyl sulfone in blood — when cooled below body temperature by being withdrawn in a syringe — formed nice white crystals, which when viewed through blood plasma, were a dead ringer for the manila-colored crystals reported by the hospital staff.

The problem is that dimethyl sulfone wouldn't have hurt anyone, and this is where the Livermore findings have become a bit controversial. If some of the dimethyl sulfone molecules had broken down in her bloodstream, they would have combined with sulfates to form dimethyl sulfate, which is a powerful nerve gas. It produces all the same symptoms that struck the emergency room staff, with the exception of nausea. It even causes the hepatitis and pancreatitis that struck Julie Gorchynski. When the paramedics started the IV line in the ambulance, the conversion of DMSO to dimethyl sulfone was only just beginning and there would have not been any dimethyl sulfate nerve gas to affect them. But by the time the hospital staff worked on her, there was just enough to knock out those working close to the drawn blood, which is exactly what happened.

Some chemists find this conversion of dimethyl sulfone into dimethyl sulfate to be implausible, but the Livermore researchers argue that this would have inevitably happened to at least some small amount of it. It's impossible to know for sure if this is what happened to Gloria Ramirez, because if it had, all the suspect compounds except the dimethyl sulfone would have evaporated away or broken back down into constituents that are normally found in the body, effectively covering their tracks and eluding the hazmat teams and the coroners. By November, even *People* magazine reported that the mystery of the toxic lady had indeed been solved, citing Ramirez' use of DMSO as the ultimate cause.

So there were now two pretty solid theories left standing, the sociogenic illness and the DMSO. Neither is perfect, and both have sound criticism.

We don't know for an absolutely certainty, and probably never will, what caused the tragic events on that February night in Riverside. But a review of the facts shows that the title of "toxic lady" is unfair and undeserved. There was nothing toxic about Gloria Ramirez; just an all-too-young cancer victim doing her best to stay alive at the end of a painful and horrible illness. Hopefully we learned something from her case that will prevent future injuries.

REFERENCES & FURTHER READING

Adams, C. "What's the Story on the Toxic Lady?" *The Straight Dope.* Creative Loafing Media, Inc., 22 Mar. 1996. Web. 30 Dec. 2011. <http://www.straightdope.com/columns/read/999/whats-the-story-on-the-toxic-lady>

Bartholomew, R., Wessely, S. "Protean Nature of Mass Sociogenic Illness: From Possessed Nuns to Chemical and Biological Terrorism Fears." *British Journal of Psychiatry.* 1 Jan. 2002, Volume 2002, Number 180: 300-306.

Editors. "Doctor Faults State Report On Faintings." *New York Times.* 4 Sep. 1994, Newspaper.

Gleick, E. "Solved: a Medical Puzzle." *People.* 21 Nov. 1994, Volume 42, Number 21: 107-108.

Stone, R. "Analysis of a Toxic Death." *Discover Magazine.* 1 Apr. 1995, Volume 16, Number 4.

Watson, R. "Coca-Cola Health Scare May Be Mass Sociogenic Illness." *British Medical Journal.* 17 Jul. 1999, Volume 319, Number 7203: 146.

Weir, E. "Mass Sociogenic Illness." *Canadian Medical Association Journal.* 4 Jan. 2005, Volume 172, Number 1.

36. THE GREY MAN OF BEN MACDHUI

A thin, dark phantom three times the height of a man is said to stalk this peak in the Cairngorms.

Today we're going to venture into the Scottish Highlands, to the bleak and misty summit of Ben MacDhui, the highest peak of the Cairngorms. At only 1309 meters it's hardly a giant compared to other mountain ranges, but it boasts a spectral giant of its own who lives there.

The summit of Ben MacDhui

Because of its high latitude, Ben MacDhui is well above the treeline and its rounded summit is a desolate field of windswept stones. It is as foggy as it is remote, yet none who venture there ever seem to feel quite alone. For more than a century, hillwalkers have been stalked by Am Fear Liath Mòr, Scottish Gaelic for The Big Grey Man. He's known best for his footsteps crunching in the gravel just out of sight; but for a certain unlucky few, the fog has thinned enough that they caught a glimpse. The Grey Man stands at least three times as tall as a man, and is dark and very thin. Some say he is covered in short brown hair, like a horse. But all who see him are filled with dread.

It is impossible to discuss the Grey Man of Ben MacDhui without relating its most famous account, that of Professor J. Norman Collie of University College London, expert in chemistry, Fellow of the Royal Society and the Royal Geographical Society, past president of the Alpine Club and member of the 1921 Mount Everest Committee. His scientific and mountaineering credentials were in good order. At the 1925 meeting of the Cairngorm Club, an association of hillwalkers of the Scottish mountain range, he told the following tale:

> *I was returning from the cairn on the summit in a mist when I began to think I heard something else than merely the noise of my own footsteps. Every few steps I took, I heard a crunch and then another crunch, as if someone was walking after me but taking steps three or four times the length of my own. I said to myself, this is all nonsense. I listened and heard it again but could see nothing in the mist. As I walked on and the eerie crunch, crunch sounded behind me I was seized with terror and took to my heels, staggering blindly among the boulders for four or five miles nearly down to Rothiemurchus forest. Whatever you make of it I do not know, but there is something very queer about the top of Ben MacDui and I will not go back there again myself, I know.*

Collie's story is today by far the most famous account of the Grey Man, but it was neither the first nor the most dramatic. In 1958, naturalist and mountaineer Alexander Tewnion wrote in *The Scots* magazine of what he described as the strangest experience of his life:

> *...In October 1943 I spent a ten day leave climbing alone in the Cairngorms... One afternoon, just as I reached the summit cairn of Ben MacDhui, mist swirled across the Lairig Ghru and enveloped the mountain. The atmosphere became dark and oppressive, a fierce, bitter wind whisked among the boulders, and... an odd sound echoed through the mist - a loud*

footstep, it seemed. Then another, and another... A strange shape loomed up, receded, came charging at me! Without hesitation I whipped out the revolver and fired three times at the figure. When it still came on I turned and hared down the path, reaching Glen Derry in a time that I have never bettered. You may ask was it really the Fear Laith Mhor? Frankly, I think it was.

The list of witnesses to the Grey Man is long, and includes many other experienced mountaineers. Does the Grey Man stalk the fells of the Cairngorms, or might there be some other explanation? As it turns out, Ben MacDhui does create a natural dead ringer for Am Fear Liath Mòr.

Brocken spectre.
Photo by Andrew Dunning.

Brocken spectres had probably been frightening people for thousands of years, but it wasn't until 1780 that Johann Silberschlag, a German member of the Prussian Academy of Sciences, first characterized them after observing them on top of the Brocken, the highest peak in northern Germany. Like Ben MacDhui, the Brocken is a relatively low, rounded hill often shrouded in fog. Silberschlag found that in foggy conditions with uniform water droplets and sufficiently strong sunlight penetration from a low enough angle, his own shadow was visible on the fog itself. Gently rounded summits like Brocken, Ben MacDhui, and many others often create ideal conditions for this, since you can stand facing downhill with the angle of the sun behind you matching the angle of the slope. If your shadow lands on the ground, it won't penetrate enough fog to create the effect — you want a long, deep shadow stretching off through the fog.

That effect can be striking. The Brocken spectre manifests itself as a very tall, very thin figure of a human, usually with disproportionately long legs. In many cases, a solar glory surrounds the spectre's head. Glories are circular, multiple-ringed rainbows caused by the backscatter of light when looking directly away from the light source; they can often be seen when you're in an airplane and you look straight down at the airplane's shadow. Brocken spectres can result from any light source, not just the sun. A full moon, or even a lantern or flashlight, will produce the spectre if the fog conditions are right. In particular, the sun and the full moon still produce enough light even when they themselves are not directly visible due to the thickness of the fog, such that there is no perceptible cause for the spectre. And in such cases, the appearance of tall, ghostly spectres can bewilder even the informed, scientific mind.

That the Brocken spectre is the source of at least some of the Ben MacDhui encounters is a certainty. As far back as 1791, the poet James Hogg was tending sheep on Ben Mac-Dhui, and saw the following:

> *It was a giant blackamoor, at least thirty feet high, and equally proportioned, and very near me. I was actually struck powerless with astonishment and terror.*

He fled home in a panic, and when he went back the next day to collect his sheep, the monster returned. This time Hogg experimented, removing his hat; and observed the figure do the same thing. He was satisfied that what he saw was merely his own shadow in the fog.

Although the visual sighting of the Grey Man himself is the most dramatic element of a meeting, it's not the most often reported. The majority of Grey Man encounters consist of sudden, unprovoked feelings of fear or of a presence, with nothing seen or heard. But it is the sound of footsteps that best characterize the Grey Man. Nearly all reports include this, and have, since the reports began. Two brothers heard the footsteps atop Ben MacDhui in 1904, "slurring footsteps as if someone was walking through water-saturated gravel." When they returned to the Derry Lodge they were told "That would have been the Fear Liath Mòr you heard," eleven years before Collie regaled the Cairngorm Club with his tale.

As fog thickens and thins, temperatures fluctuate, and rocks expand and contract and split. Ice also splits rocks. When either of these happen on a slope, a rock may tumble. These actions are, in fact, entirely responsible for the crumbled stone of which Ben MacDhui consists. Even on a calm day, rocks make such noises, everywhere.

Animals such as deer and wildcats abound on Ben Mac-Dhui; the reason Alexander Tewnion had a pistol on his 1943 trek was to shoot hares and ptarmigan. Hikers disturb nearby creatures, and hiking anywhere will always produce the sounds of some scattering animal. There's no doubt that these sources of sound account for at least some of the noises reported by Grey Man witnesses.

Of course, suggesting possible sources for the sounds perceived as following footsteps does not prove that something unknown wasn't actually following some of these hikers. If it was, it remains unknown, and does not constitute evidence for a tall, hairy man-beast any more than it constitutes evidence for a spritely leprechaun. What's needed is testable evidence; and as it turns out, there has been one case of the Grey Man's giant footprints being discovered in the snow, and photographed.

The footprints were seen and the photographs published by John A. Rennie in his book *Romantic Strathspey,* and the prints he found were in the Spey Valley some 15 miles away from Ben MacDhui. The prints:

> *...were running across a stretch of snow covered moor-land, each print 19 inches long by about 14 inches wide*

*and there must have been all of seven feet between each
"stride". There was no differentiation between a left
and a right foot, and they preceded in an approximate-
ly single line.*

And that's where the popular tellings of the Grey Man's
footprints end; only with the observation, and not with the ob-
server's own explanation. For, on a second occasion, Rennie
saw such prints again, and this time he watched them form in
the snow. They were the result of precipitation. And in his own
words, Rennie said of the footprints:

*In that moment I knew that the Wendygo, Abominable
Snowman, Bodach Mor, or what have you, was forever explained so far as I was concerned.*

Thus we are left with no evidence that the physical feet of
an unknown creature have ever created the sounds of footsteps
on Ben MacDhui.

Some of the other most often cited evidence of the Grey
Man also weakens under scrutiny. Although Norman Collie's
story is widely considered the most authoritative, it should be
taken with a grain of salt within the context of Collie's personality. According to his biographer Christine Mill, he was a lifelong believer in the occult. In later years after 1933 when the
Loch Ness Monster became a phenomenon, he also became a
firm believer in that; which, like the Grey Man, he never saw
(it's an often-overlooked point of Collie's famous story that he
did not report seeing anything; he only heard footsteps and felt
frightened). Mill wrote that Collie would often tell stories
around the campfire or in his den of Gaelic mountain gods and
goddesses, and other legendary creatures; and as she put it, "No
one quite knowing how much he was believing himself." Unfortunately, nothing about Collie's account raises its status from
that of campfire story to that of useful evidence.

And the realm of campfire stories is perhaps where the Grey Man of Ben MacDhui best belongs. Never underestimate the power of nature to magnify feelings of dread, loneliness, and isolation. Everyone who has spent a night in a tent outdoors knows the effects of small, unexplained sounds on the mind. The cold, foggy Cairngorms are the ideal place for such sensations to augment, and for our animal senses to trigger animal responses. The Grey Man need not be a physical creature for it to be — as far as our minds are concerned — utterly real.

REFERENCES & FURTHER READING

Collie, J. *Annual General Meeting of the Cairngorm Club.* Aberdeen: Cairngorm Club, 1925.

Gray, A. *The Big Grey Man of Ben Macdhui.* Aberdeen: Impulse Publications, 1970. 39-40.

Hastie, J. "Big Grey Man - The Evidence." *Scottish Mountaineering Club Journal.* 1 Jan. 1998, Volume 36, Number 189: 507-513.

Hazen, H. "The Brocken Spectre." *Popular Science.* 31 Dec. 1900, Volume 34, Number 6: 106-107.

Kaczynski, R. *Perdurabo: The Life of Aleister Crowley.* Tempe: New Falcon Publications, 2002. 31.

Rennie, J. *Romantic Strathspey: Its Lands, Clans and Legends.* London: R. Hale, 1956. 83.

Roberts, A. "The Big Grey Man of Ben MacDhui and Other Mountain Panics." *Fortean Studies.* 1 Jan. 1998, Volume 5.

Ross, H. *Behaviour and Perception in Strange Environments.* London: George Allen & Unwin, 1974. 54-56.

Tewnion, A. "A Shot in the Mist." *The Scots Magazine.* 1 Jun. 1958, Volume 69: 226-227.

37. Wunderwaffen: Nazi Wonder Weapons

Did the Nazis really have super-advanced technology, even anti-gravity flying saucers?

Few subjects provoke as much emotion as Nazi Germany, or attract as much attention and speculation. Since the war, we've even attached an occult mythology to Nazism, in an attempt to rationalize it away as having come from outside of our own society. This combination of true military might and mysticism has spawned a whole subculture of study of Nazi *Wunderwaffen*, the alleged wonder weapons with capabilities that far exceeded those of the Allied forces not only of the 1940s, but even of today. The range of these weapons goes from simple gunsights to ramjet fighter planes, and even all the way to antigravity flying saucers. How much of this mythology is true, and how much is driven more by our fascination with occultifying Nazism?

Like all military industrial complexes, Nazi Germany had military research programs, as did a huge number of civilian contractors. Within all of these scores of programs, serious plans for just about any advanced weapon you can imagine did in fact exist. As Germany's resources and manpower dwindled over the course of the war, fewer and fewer of these projects saw the light of day, but some of those that did were astonishingly advanced for their time.

We know about virtually everything that was under development in Nazi Germany because at the war's end, the Allied forces overran Germany and captured not only all of their technology in the form of operational and prototype designs, but also all of the documentation pertaining to their experiments and plans. In many cases, documentation was destroyed by the Nazis as capture became imminent; but this primarily regarded activities that were likely to be prosecuted as war crimes, such as the human experimentation programs at places like Auschwitz. All of the significant factories and design bureaus were captured relatively intact, and we have a very complete picture of what the Nazis did and did not develop.

Me-163 *Komet*

Real weapons that the Nazis did actually build and deploy included jet powered fighters such as the Messerschmitt Me-262 and Heinkel He-162, and even a rocket powered fighter, the Me-163. There were also a number of variants and derivatives of these and similar aircraft. Toward the end of the war, some troops were armed with the Zielgerät ZG-1229 *Vampir* infrared gun sight, giving them night vision years before most Americans had ever dreamed of such a thing. Perhaps the pinnacle of Nazi military might was the pulsejet powered V-1 guided cruise missile, and the suborbital V-2 long-range ballistic missile, three thousand of which entered space fifteen years before Sputnik 1.

Other designs, while seemingly even more fanciful, did in fact exist, in either prototype form or completed (and perfectly sound) blueprints. Aircraft included the Horten Ho-229 jet powered flying wing, the Mach 2.2 Lippisch P.13a delta winged ramjet-powered fighter, a high altitude spyplane similar to the later American U-2 called the DFS 228, even a variable geometry swing-wing jet, the Messerschmitt P.1101, which became the precursor to the later American Bell X-5. They also had designs for a number of vertical takeoff and landing jets.

The Nazis also aggressively pursued their *Amerika Bomber* program, hoping to create a system with the range to bomb the United States from Germany. These included variants of the Arado E.555 jet powered flying wing, and even a suborbital spaceplane called the *Silbervogel* which went as far as a glide test mockup. There were many, many other candidates for *Amerika Bombers* as well.

On land, the Nazis had plans for a pair of staggeringly gigantic tanks, the Landkreuzer P.1000 *Ratte* and P.1500 *Monster*, crewed with over 40 and 100 men respectively. The P.1500 would have fired the largest artillery projectiles ever designed, the 800mm railroad gun.

At sea, the Nazis planned to equip a new type of U-Boat to fire their V-2 missiles into the United States, called the Rocket U-Boat. Three were ordered, and one was actually built, thought its testing was not completed before the war's end. And what would it have carried?

Nazi atomic warheads atop V-2 missiles were nearer to a reality than most people realize. While the Manhattan Project was happening in the United States, it had a twin hard at work in Germany: the *Uranverein*, or Uranium Club. The Uranium Club had just as strong a start as the Manhattan Project, perhaps even stronger; but Germany's rapidly diminishing resources over the course of the war meant that it couldn't be as fully staffed or funded as was the Manhattan Project. The operation of Germany's reactors for the breeding of plutonium required heavy water, which came from the Vemork hydroelectric plant in Norway, originally built to produce nitrogen for agriculture. The final nail in the Uranium Club's coffin came from perhaps the most important sabotage job in history: Operation Gunnerside, in which a small team of Norwegian commandos were airdropped and skied to Vemork. They climbed the cliffs surrounding the plant, entered through a utility duct, and planted explosives around the electrolysis chambers. The resulting explosions destroyed Germany's entire supply of heavy water and most of the equipment needed to produce it. 3,000 troops were sent after them, but the Norwegian commandos all escaped.

Several months later production resumed, but was hampered by severe allied bombing. Germany attempted to deliver what heavy water it had, and put the casks on a ferry. One of the commandos, Knut Haukelid, was in the area and managed to plant a bomb on board the ferry, which sank in deep water. This marked the end of Nazi Germany's atomic weapons program.

Several authors have alleged that Uranium Club scientists did, on several occasions, actually test atomic bombs. These were either hollow cores — meaning the shaped charge was in place to implode the plutonium core, but there was no plutonium — or a paraffin or silver core seeded with deuterium. But science historians have doubted the at-best controversial evidence supporting these claims, and Germany's Federal Physical and Technical Institute performed soil tests in 2006 where the tests are said to have happened and failed to find any chemical signatures.

And all of this brings us to the final, and most incredible, of the Nazi *Wunderwaffen*, known as *Die Glocke*, which means the Bell. The Bell is said to have been a saucer shaped aircraft, usually powered by a pair of rotating drums containing a mysterious iridescent purple liquid. It is this family of Nazi flying saucers, known by various nicknames and designations, that most of the *Wunderwaffe* mythology centers around. In all of the data and materials captured by the occupying forces, nothing remotely like the Bell was ever discovered, alluded to, or even imagined. There is, quite simply, no record indicating that anything like it existed, outside of the undocumented claims made by a number of authors and individuals decades later.

The inspiration of nearly everything found on the Internet today about Nazi flying saucers is a book, written in the year 2000 by Polish military historian Igor Witkowski called *The Truth About The Wunderwaffe*. Witkowski told an amazing tale: He was given access to (but not allowed to copy) the classified transcript of an interrogation by Polish agents of the Nazi SS officer Jakob Sporrenberg. Through this transcript, Witkowski claimed to have learned about *Die Glocke*. This account became popular in the West when aviation writer Nick Cook included it in his popular 2002 book *The Hunt for Zero Point*, a tale of the cranks and colorful characters who have tried to invent anti-gravity machines. Since that time, you've been able to find all you want on the Internet about Nazi flying saucers.

Whether Witkowski actually saw such a transcript, or just made it up, is unknown. He offered no evidence of its existence and nobody else, inside or outside of Poland, has ever reported seeing such a thing. But what is known is that the SS officer Sporrenberg can't corroborate Witkowski's claim. Sporrenberg was executed as a war criminal in 1952. He'd been a field officer fighting partisans, and had never had any connection with science or aviation branches of the Nazi military.

Hoax photo of *Die Glocke*

But there had been an existing mythos to anchor Witkowski's Glocke. Mythology has always surrounded the Nazis. Perhaps because of how incomprehensible was the Holocaust, post-war fascination with Nazism has tried to explain it away as the result of some demonic influence stemming from mysticism and occultism. The Nazi regime has always been a magnet for occult theories. This was born mainly in 1960 when two French authors wrote a fanciful work called *The Morning of the Magicians* in which they speculated about many mystical communities in Germany, among which was one inside pre-war Berlin called the Vril Society. The secretive Vril Society was said to be an inner circle among inner circles of various mystical, New Age, and occult orders. The book claimed the Vril Society formed the nucleus of the Nazi party. No reference to a Vril Society has been found documented prior to this book.

But the mysterious substance Vril was itself already embedded in popular consciousness. It had been since 1870 when the popular English writer Edward Bulwer-Lytton published a science fiction novel called *The Power of the Coming Race*. In this story, the population of Atlantis escaped their sinking nation by fleeing to the hollow center of the Earth. They possessed a magical fluid called Vril, which served as a limitless power source and the elixir of life.

I've been able to find only one thread linking Bulwer-Lytton's fanciful novel to the Nazis. In 1935, German astronomer and rocket scientist Willy Ley emigrated to the United States, as did many of his countrymen. Ley was also a prolific writer, and mixed science fiction in with his science writing. For *Astounding Science Fiction* he wrote an article called "Pseudoscience in Naziland" in which he described a group that was:

> ...*literally founded upon a novel. That group which I think called itself Wahrheitsgesellschaft — Society for Truth — and which was more or less localized in Berlin, devoted its spare time looking for Vril.*

And so we have a more-or-less complete timeline of the genesis of the Nazi UFOs. They are entirely the invention of authors outside of Germany, leveraging the public's hunger for strangeness associated with the Nazis. Today, any Internet search for some of these terms will yield a tsunami of hoaxed black-and-white photographs, conspiracy theories of cover-ups, interviews with cranks claiming to have some insider knowledge, and endless lists of model numbers and designations of Nazi flying saucers that never existed. Within aviation and military history, no reference exists to flying saucers powered by drums of Vril or antigravity technologies.

It is the very nature of our perception of the Nazis that drives these tall tales of *Wunderwaffen*, not actual history. It's another case where the real wonder is in *why* the legend exists, not the legend itself. The Bell might never have flown, but it still offers us a fascinating lesson on why we believe.

References & Further Reading

Bulwyer-Lytton, E. *The Coming Race.* Edinburgh: W. Blackwood and Sons, 1871.

Cook, N. *The Hunt for Zero Point.* New York: Broadway Books, 2002.

Cornwell, J. *Hitler's Scientists: Science, War, and the Devil's Pact.* New York: Viking, 2003.

Forsyth, R. *Messerschmitt Me-264 Amerika Bomber: The Luftwaffe's Lost Transatlantic Bomber.* Hersham: Classic, 2006.

Gallagher, T. *Assault in Norway: Sabotaging the Nazi Nuclear Program.* Guilford: The Lyons Press, 2002.

Jane's. *Jane's Fighting Aircraft of Word War II.* London: Bracken Books, 1989.

Karlsch, R., Walker, M. "New Light on Hitler's Bomb." *Physics World.* Institute of Physics, 1 Jun. 2005. Web. 15 Jan. 2011. <http://physicsworld.com/cws/article/print/22270>

Pauwels, L., Bergier, J. *The Morning of the Magicians.* New York: Stein and Day, 1963.

Witkowski, I. *Prawda o Wunderwaffe.* Warszawa: WiS-2, 2002.

38. FINDING AMELIA EARHART

Popular modern reports claim Amelia Earhart made it to an island and survived for a time. Might that be true?

Today we're going to point the skeptical eye at some of the rumors surrounding one of the twentieth century's great mysteries: The disappearance of pioneering woman aviator Amelia Earhart, when her airplane disappeared over the Pacific Ocean

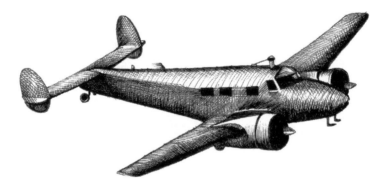

Lockheed Model 10 *Electra*

on her famous 1937 flight around the world. Conventional wisdom says that she simply ran out of fuel and ditched into the ocean, but stories have persisted for decades that she might have made it safely to an island, perhaps even survived for some time. Here and there, various artifacts have been found: A shoe, a zipper, a scrap of aluminum. There are even some crazy stories: that she made it back to the United States and lived out her life under an assumed name, or that she was captured by the Japanese and executed as a spy. Let's take a look to see if any of these alternate explanations can withstand scrutiny.

Amelia Earhart and her navigator, the highly experienced and esteemed Fred Noonan, were on the third-to-last leg of their circumnavigating flight in her Lockheed Electra 10E, a 1200 horsepower, state-of-the-art twin engine aircraft. They took off from Lae in Papua New Guinea on July 2, 1937, headed for a remote refueling stop in the South Pacific, a tiny island called Howland. From there they would continue to Honolulu for a final refueling before completing the journey in Oakland, California.

And as everyone knows, they never made it to Howland. A US Coast Guard cutter, the *Itasca*, was on station at Howland transmitting a radio direction-finding signal, and made sporadic voice contact. Most historians agree that a half-hour time zone difference disrupted both parties' attempts to establish two-way voice communication, and a photograph of the Electra taking off from Lae appears to show that a belly antenna (of unconfirmed purpose) may not have been in place. And to top it off, it turns out that Howland's position was misplaced on Earhart's chart by about five nautical miles, but which would still have kept it within visual range. Whatever role these problems may have played, if any, is unknown; but Earhart's final radio transmission to the *Itasca* said they were in the immediate vicinity of Howland. And ever since then, the best analysis is that they ran out of fuel, ditched in the Pacific Ocean, and perished.

But one group of historic aviation enthusiasts called TIGHAR (The International Group for Historic Aircraft Recovery) has been tirelessly promoting their hypothesis that Earhart and Noonan flew not to Howland, but by mistake to an island 650 km to the southeast, now called Nikumaroro but then called Gardner, where they crashed and survived for a time as castaways. TIGHAR's hypothesis and claimed discoveries saturate virtually all television and print reports of Earhart for the past decade, but these media outlets almost never mention that TIGHAR's is a fringe theory supported by poor evidence and that has almost no serious support from mainstream historians or archaeologists.

Here's the problem with TIGHAR's findings. Even though they meticulously document and preserve every artifact, they exhaustively research each one to find matches with real objects from the 1930s, and they look exactly like what such an expedition should look like, their overall methodology is fundamentally, fatally unscientific. It's unscientific in that it's done completely backwards. TIGHAR begins with the assumption that Amelia Earhart crashed, camped out, and died on Nikumaroro. They take everything they find — every anomaly in a photograph or in a story, every piece of bone or manmade artifact found on the island — and try to match it to their assumption, rather than trying to objectively assess its origin.

Nikumaroro, this tiny island where TIGHAR has recovered its artifacts, is in Kiribati, a nation of 100,000 people spread over millions of square kilometers of the South Pacific. People leaving artifacts come and go all the time. For example, pearl divers. Fleets of pearl boats have plied these waters since the 1800s. Every island and reef in the South Pacific has been visited countless times by pearl boats, who anchored, made camp on shore, and spent a few weeks free diving for oysters. Their exploits and histories have been published in dozens of books, such as Roy Miner's 1941 volume *Pearl Divers,* and the many colorful tales in Frank Coffee's 1920 book *Forty Years on the Pacific.* TIGHAR found evidence of campfires and fish bones on Nikumaroro and concluded "Amelia Earhart" who is not known to have visited the island; but I found no attempt made by them to exclude the pearl divers who *are* known to have camped there, and to have done so countless times over more than a century. TIGHAR appears to be dedicated to proving the *least* likely explanation for the artifacts.

They found an object identified as the heel of a woman's shoe. Many pearl divers were women, and they came from Fiji, the Philippines, and New Zealand, where shoes were not unknown in the 1930s. Could the shoe have come from the 1929 wreck of the steamship SS *Norwich City* that killed 11 of its 35 crew on Nikumaroro? Could it have belonged to one of the sixteen women who settled on the island in 1939 as part of a British colony? Could the shoe have even floated to the island from anywhere else? I find no reason to exclude the women who lived on or visited the island as possible owners of the shoe, or any reason to suggest Amelia Earhart was the most likely owner.

The found the remains of a buck knife. Is Amelia Earhart really more likely to have brought a buck knife to Nikumaroro than pearl divers, the British settlers, the operators of an 1892 coconut plantation, or the 25 crew of a 1944 Coast Guard station?

At its height, Nikumaroro had a population of about 100 people. Half a dozen smaller populations had come and gone over the prior century, and throughout it all, pearl divers camped ashore. Would you expect such an island to be pristine, or would you expect random debris from not just the 1930s but other periods as well? Without exception, every one of the artifacts recovered by TIGHAR should be expected to have been found there whether or not Amelia Earhart had ever even lived.

This even extends to a partial human skeleton that was found on the island in 1940 during its British colonial occupation. At the time, the young officer who found it, Gerald Gallagher, shipped the bones to Dr. David Hoodless, principal of the Central Medical School of the South Pacific on Fiji. Hoodless studied the bones and reported them to be "definitely" male, judging by the pelvis; and from an individual about 5 foot 5 1/2 inches tall, of European heritage and not a Pacific Islander. No clothes or hair were found, and the bones were severely weatherbeaten and in poor condition.

Near the skeleton, Gallagher also found a small wooden box with dovetailed joints, that he determined to be a sextant box. It was delivered to Harold Gatty, founder of Air Pacific, and a good friend of Fred Noonan and familiar with his navigation habits; for example, that he often carried an old-school sextant with him on flights in addition to modern equipment, just to double-check things the way a good navigator should. Regarding Gatty's own expertise, Charles Lindbergh had described him as the "prince of navigators". Another British officer in the area cabled Gatty's findings back to Gallagher:

> *Mr. Gatty thinks that the box is an English one of some age and judges that it was used latterly merely as a receptacle. He does not consider that it could in any circumstance have been a sextant box used in modern trans-Pacific aviation.*

After studying all these results in light of his original speculation that they may have been related to Earhart, Gallagher wrote:

> *It does look as if the skeleton was that of some unfortunate native castaway and the sextant box and other curious articles found nearby the remains are quite possibly a few of his precious possessions which he managed to save.*

Neither the bones nor the sextant box still exist today, but TIGHAR has made their own analysis of them, based on reading these original reports. As expected, TIGHAR has concluded that the skeleton *was* consistent with that of Amelia Earhart, and that the sextant box *was* consistent with one Fred Noonan may have used. Essentially, TIGHAR took the original first-hand expert analyses, and rejected and re-interpreted them to support their desired conclusion.

From a navigational perspective, the fundamental assumption of TIGHAR's theory is almost inconceivable. Fred Noonan was one of aviation's top experts in using the latest navigational techniques and equipment, including the then-new E-6B flight computer, which (among other things) corrects for the effects of wind on speed and course. Nikumaroro is a full five and one half degrees of latitude south of Howland. That's a massive, massive error; it's simply not plausible that Noonan could have been that far wrong. Earhart was no slouch of a navigator either. Could they have made such an error without either of them catching it?

Moreover, the bearing from Papua New Guinea to Howland is about 79° true. To Nikumaroro, it's 89° true. Nikumaroro was about 4272 km away, only slightly farther than Howland, which was 4160 km. The Electra's maximum's range did allow them to make it to either island, but only if they flew an absolutely direct course. The TIGHAR hypothesis suggests that they made their entire flight at a full 10° off course, without catching it, while following their compass and homing in on the *Itasca's* direction-finding signal, and were as much as five degrees of latitude too far south. Even for 1937, this size of an error strains credibility. Either the E-6B or the sextant would have caught either of these errors easily.

Howland Island, the intended destination, is basically just a flat coral sand cay in the middle of nowhere, about two and a half kilometers long and less than a kilometer wide. It's uninhabited and has no trees, and no structures other than an automated lighthouse beacon. It's about as featureless and bleak as a desert island can be. But in 1937, there was a tiny temporary population there. Hawaii's Kamehameha School for Boys had set up a camp where students would spend a few months learning about the plants and animals there. It was called Itascatown, named after the *Itasca* that supplied it and handled all the transportation of students.

Three unpaved runways were bulldozed in anticipation of Earhart's landing, but since she never arrived, they ended up having never been used at all. The Japanese bombed them during World War II and they were never repaired.

But back on that day in 1937, the airstrips were ready, the *Itasca* sat on station off the coast of Howland, and drums of fuel had been sent ashore to refuel Earhart's plane. Coast Guardsmen and teenagers from the Kamehameha School stood watching the skies. They watched and waited, the time for Earhart's arrival came and went, and still they watched. The skies remained quiet. Eventually it became clear that there would be no landing that day, and word gradually spread that the *Itasca* had lost contact and the plane was now well past the point at which its fuel would have run out.

Following the bearings of Earhart's final radio transmission, just northwest of Howland, the search ships combed the ocean for a week. The aircraft carrier USS *Lexington*, the battleship *Colorado*, the *Itasca*, and even a few Japanese ships scoured the ocean's surface, tiny gray dots on an unimaginably vast shimmering blue curtain. But well hidden, deep in the peaceful darkness thousands of fathoms below them, rested what remains aviation's most enduring legend.

REFERENCES & FURTHER READING

Coffee, F. *Forty Years on the Pacific.* New York: Oceanic Publishing Company, 1920.

Editors. "Sextant Box Found on Nikumaroro." *Earhart Project Wiki.* TIGHAR, 25 Jan. 2010. Web. 27 Jan. 2012.
<http://tighar.org/wiki/Sextant_box_found_on_Nikumaroro>

Goldstein, D., Dillon, K. *Amelia: The Centennial Biography of an Aviation Pioneer.* Washington, D.C.: Brassey's, 1997. 245-254.

Leff, D. *Uncle Sam's Pacific Islets.* Stanford University: Stanford University Press, 1940. 47-50.

Lorenzi, R. "Amelia Earhart Clue Found in Clumps." *Discovery News.* Discovery Communications LLC, 2 Mar. 2011. Web. 27 Jan. 2012. <http://news.discovery.com/history/amelia-earhart-clumps-island-castaway-clues-110302.html>

Miner, R. *Pearl Divers.* New York: American Museum of Natural History, 1941.

Strippel, R. "Researching Amelia: A Detailed Summary for the Serious Researcher into the Disappearance of Amelia Earhart." *Air Classics.* 1 Oct. 1995, Volume 31, Number 11: 20.

USCG. "What was the Coast Guard's role in the search for Amelia Earhart?" *Coast Guard History.* United States Coast Guard, 10 Oct. 2012. Web. 15 Nov. 2012.
<http://www.uscg.mil/history/faqs/earhart.asp>

39. THE VERSAILLES TIME SLIP

Did two women visiting Versailles in 1901 time-travel back to the days of Marie Antoinette?

The year was 1901, and a pair of friends, 55-year-old Anne Moberly and 38-year-old Eleanor Jourdain, were on holiday in France. They were both teachers at St. Hugh's College in Oxford. Moberly was in fact the Principal there, and Jourdain would become her successor fourteen years later. With a Baedeker's tourist guidebook in hand, the two set out to see the vast Palace of Versailles, the center of political power in France until the French Revolution in 1789. They turned to visit the *Petit Trianon,* a small chateau on the grounds given by Louis XVI to his 19-year-old wife, Marie Antoinette, as a private retreat for her personal use.

Moberly and Jourdain got a bit lost searching for the chateau, and it was during this interlude that they made history, even if only in some small way. They encountered several people in 1789 period attire, carrying out period activities, and passed a handful of structures that had not existed since 1789. Their unexpected visit to 112 years in the past culminated with an encounter with Marie Antoinette herself, sketching on the grounds of her chateau. It was only upon being snatched out of their reverie by a modern tour guide that Moberly and Jourdain suddenly found themselves once again in 1901.

Within months they'd published their account in a book called *An Adventure,* and as they were both respected academics who did not desire bizarre publicity, they published it under pseudonyms. Their experience became variously known as the Versailles Time Slip, the Ghosts of Trianon, or the Moberly-

Jourdain Incident; and it has intrigued researchers, historians, and enthusiasts of the paranormal ever since.

My own interest in this tale was triggered by the evident reliability of the witnesses. In so many cases of strange and inexplicable phenomena being reported, we hear that the witness or witnesses were people of great character or scientific credibility, and so it's often considered implausible that they could be mistaken or making the story up. The Versailles Time Slip is clearly one such case. Must we accept that the event happened as reported, due to the credibility of Moberly and Jourdain? Whether you've heard of the story before or not, it's a good lesson in how we should best assess cases where the witnesses' credibility is offered as evidence that the event was factual.

Author Kathleen McGowan described the ladies as:

> *Two highly educated Englishwomen women of impeccable reputation...*

Writer Nell Rose observed:

> *They were not liars, and both ladies had nothing to gain by making up this story. In fact it could go a long way to ruining their reputation.*

Frequently I'll hear something along the lines of "My Uncle Bob was a very trustworthy and honest man who would never make something up, therefore you should accept his ghost experience as a fact." Moberly and Jourdain's position as college teachers also gilds them with a cloak of authority, similar to that given to pilots or astronauts who see UFOs. Certainly a pilot's perception of a UFO cannot be mistaken, and certainly college teachers' perception of going back in time must have therefore happened as they thought. Well, not so much. It's not necessary to suggest that Reliable Uncle Bob must have been a liar for either his own perception to have been mistaken, or for the version of the story that finally made it down to you to have been altered. Neither is it necessary to cast doubt on Moberly and Jourdain's academic status to suggest that the story we have in the annals of urban legend does not reflect a literal time-travel event that must have happened exactly as reported. Human brains are fallible — including Uncle Bob, including pilots, and including academics.

Two basic explanations have been put forth by previous researchers, which by now have been watered down and popularized into the following: First, that they accidentally wandered into a historical reenactment; and second, that they had a sort of shared delusion. But to put these two explanations into proper perspective, we need to go back to see who originally proposed them and why, and what hidden details of the story prompted them. And this is the point at which the Versailles Time Slip goes from an interesting anecdote to an all-out strange-fest.

St. Hugh's College was founded in 1886 by Elizabeth Wordsworth as an all-women's college at Oxford, and the consensus among researchers suggests that Moberly and Jourdain's relationship was romantic as well as professional. They were at least roommates. One of the earliest and most popular critiques of *An Adventure* came in 1957 from their former student Lucille Iremonger in her prodigiously-entitled book *Ghosts of Versailles: Miss Moberly and Miss Jourdain and Their Adventure: A Critical Study.* Iremonger insinuated that both women even had frequent affairs with students of the college. She spent much time delving into the nature of their lesbian relationship, and basically concluded that their adventure was a *folie a deux,* a madness of two. They had, she suggested, been so distracted by their relationship that they had merely misinterpreted ordinary people and objects for things from 1789, and became so obsessed with proving their story that they'd even convinced themselves of the reality of what had happened. Iremonger's charge gradually became softened over the years into a "shared delusion".

A fourth edition of *An Adventure* was printed in 1955, this time with a preface written by art historian Joan Evans, who was Jourdain's literary executor, and like Iremonger, a former student. But unlike the hostile Iremonger, Evans tended to defend Moberly and Jourdain's account of what happened as a literal event. She felt compelled to deflect popular conjecture that they'd had some sort of strange lesbian romance-induced delusion. So Evans, in a 1976 article for *Encounter* magazine, put forth the suggestion that the two women had simply walked unknowingly into an historical recreation, in which actors were lounging about in period attire. Evans went so far as to research such recreations, but did not find such an event that would have coincided with the 1901 visit.

Evans turned to the 1965 biography of the French artist Robert de Montesquiou. Biographer Philippe Jullian noted that de Montesquiou had lived in a house at Versailles and was noted for his Tableaux Vivant performances, in which gay Parisian men performed the roles of both men and women; thus, the Marie Antoinette seen by Moberly and Jourdain was a transvestite. Though no evidence survives that indicates de Montesquiou may have actually thrown such an event in 1901,

Jullian's suggestion was good enough for Evans; and ever since her article, the transvestite historical recreation has been reported and re-reported as one of the most likely explanations for the Versailles Time Slip. They say truth is stranger than fiction, but the rationales for fringe claims can often be even stranger.

As usual, the best way to get a handle on what probably actually happened is to brush aside all *ex post facto* conjecture — the lesbian madness and transvestite follies — and go back to the original sources. One thing I like to do, since I've never visited Versailles, is to pull it up on Google Earth and look at all the pictures of it I can find. The first thing one finds is that the grounds of Versailles are immense, about 3.5 kilometers from end to end. To get to the *Petit Trianon,* you have to cross whole square kilometers of gardens, lakes, little hamlets and chateaus. Part of Moberly and Jourdain's proof is that upon returning a few years later, they couldn't find some of the objects they'd witnessed, most notably a kiosk and a footbridge. When they sent their story to England's Society for Psychical Research, a sometimes-skeptical, sometimes-credulous association of enthusiasts of the paranormal, the Society was unimpressed. Part of what the Society noted was that Moberly and Jourdain had themselves stated that they'd been lost; and as footbridges and kiosks of various descriptions abound on the vast grounds of Versailles, there was almost nothing to go on and nothing surprising about their report.

In their published 1950 report of what they'd determined many years before, the Society noted a few other points that authors like Iremonger tore into like fresh meat. One was that when they reviewed the several editions of *An Adventure,* they found it had expanded notably each time. Moreover, it was three months after the incident before the women had even sat down to compare notes on what they'd witnessed; whereas *at the time of their visit, neither woman had suspected anything unusual even took place!* In the second edition of *An Adventure,* the women explained that a full three months after their visit to Versailles, Moberly happened to mention the sketching woman they saw. Jourdain didn't remember any such thing. As they talked, it turned out that Moberly didn't remember hardly anything that Jourdain did either. These were all minor details like a woman shaking out a cloth out a window, two green-jacketed gardeners at work, and a sinister-looking man sitting under a garden kiosk. It was only after much discussion, note-sharing, and historical research that Moberly and Jourdain came up with the time period as 1789 and assigned identities to a few of the characters they saw, including Marie Antoinette herself as the lady sketching on the lawn.

Upon reviewing the case as told in the women's own words, the Society for Psychical Research concluded that the evidence of anything unusual having actually happened was insufficient to justify further study. Their report's author, W.H. Salter, pointed out that the embellished versions of the tale published in later editions were written much later than the women had initially claimed, perhaps as long as five years later; and only after the women had made several return trips to Versailles to study the landmarks further.

The principal authors who have written about this story seem to agree that there was probably no conscious effort at deception by Moberly and Jourdain, only a firm belief in the reality of their perception and a desire to present their story in as convincing a way as possible. They even went so far as to include a chapter they called "A Rêverie", an imaginary account of Marie Antoinette's own meditations, in which she observed two strangers walk past while she was sitting there sketching, amidst all the other people and things they reported. This chapter jumps out as particularly bizarre, and moves *An Adventure* from the realm of reporting into that of fantasy fiction.

But it was probably firmly believed by both women. In paraphrasing Iremonger, author Terry Castle wrote:

> *...Both Moberly and Jourdain had had paranormal experiences before and after the Trianon visit, and that Moberly in particular was prone to aural and visual*

hallucinations. As a child she had heard the words "PINNACLED REALITY" as she stared at the spires of Winchester Cathedral; ...she had seen two strange birds with dazzling white feathers and immense wings fly over the cathedral into the west. In Cambridge in 1913 she saw a procession of medieval monks; and at the Louvre the following year, she saw a man "six or seven feet high" in a crown and togalike dress whom she at first took to be Charlemagne, but later decided was an apparition of the Roman emperor Constantine.

Respected academics they may well have been, and well-intentioned to boot. But no one is above being mistaken, or above susceptibility to preconceived notions and all manner of perceptual errors. I'm not, you're not, Uncle Bob is not, and Moberly and Jourdain were not. Honesty, integrity, credentials, and respect have nothing to do with the human brain's function of abstracting its perceptions into something that seems to make sense. There was never any need for authors to introduce lesbian madness and transvestite follies to explain their erroneous perception. Moberly and Jourdain were simply human; and that, in itself, is the most complex explanation of all.

REFERENCES & FURTHER READING

Boese, A. "The Ghosts of Versailles." *Museum of Hoaxes.* Alex Boese, 11 Apr. 2001. Web. 3 Feb. 2012.
<http://www.museumofhoaxes.com/versailles.html>

Castle, T. "Contagious Folly: An Adventure and Its Skeptics." *Critical Inquiry.* 1 Jul. 1991, Volume 17, Number 4: 741-772.

Evans, J. "End to An Adventure: Solving the Mystery of the Trianon." *Encounter.* 1 Oct. 1976, Volume 47, Number 4: 33-47.

Iremonger, L. *Ghosts of Versailles: Miss Moberly and Miss Jourdain and Their Adventure: A Critical Study.* London: Faber and Faber, 1957.

Jullian, P. *Robert de Montesquiou, un Prince 1900.* Paris: Librairie Acadeℕmique Perrin, 1965.

Morison, E., Lamont, F. *An Adventure.* London: MacMillan and Co., 1901.

Salter, W.H. *An Adventure: A Note on the Evidence.* London: Society for Psychical Research, 1950.

40. A Magical Journey through Reasoning Errors

Four common types of analytical errors in reasoning that we all need to beware of.

Today we're going to cover a bit of new ground in the basics of critical thinking and critical reasoning. There are several defined types of common analytical errors to which we're all prone; some, perhaps, more so than others. Reasoning errors can be made accidentally, and some can even be made deliberately as a way to influence the acceptance of ideas. We're going to take a close look at the Type I false positive error, the Type II false negative error, the Type III error of answering the wrong question, and finally the dreaded Type IV error of asking the wrong question.

By way of example we'll apply these errors to three hypothetical situations, all of which should be familiar to fans of scientific skepticism:

- From the realm of the paranormal, a house is reported to be haunted. The null hypothesis is that there is no ghost, until we find evidence that there is.

- The conspiracy theory that the government is building prison camps in which to orderly dispose of millions of law-abiding citizens. The null hypothesis is that there are no such camps, until we find evidence of them.

- And from alternative medicine, the claim that vitamins can cure cancer. The null hypothesis is that they don't, unless it can be proven through controlled testing.

So let's begin with:

Type I Error: False Positive

A false positive is failing to believe the truth, or more formally, the rejection of a true null hypothesis — it turns out there's nothing there, but you conclude that there is. In cases where the null hypothesis does turn out to be true, a Type I error incorrectly rejects it in favor of a conclusion that the new claim is true. A Type I error occurs only when the conclusion that's made is faulty, based on either bad evidence, misinterpreted evidence, an error in analysis, or any number of factors.

In the haunted house, Type I errors are those that occur when the house is not, in fact, haunted; but the investigators erroneously find that it is. They may record an unexplained sound and wrongly consider that to be proof of a ghost, or they may collect eyewitness anecdotes and wrongly consider them to be evidence, or they may have a strange feeling and wrongly reject all other possible causes for it.

The conspiracy theorist commits a Type I error when the government is not, in fact, building prison camps to exterminate citizens, but he comes across something that makes him reject that null hypothesis and conclude that it's happening after all. Perhaps he sees unmarked cars parked outside a fenced lot that has no other apparent purpose, and wrongly considers that to be unambiguous proof, or perhaps he watches enough YouTube videos and decides that so many other conspiracy theorists can't be all wrong. Perhaps he simply hates the government, so he automatically accepts any suggestion of their evildoing.

Finally, the alternative medicine hopeful commits a Type I error when he concludes that vitamins successfully treat a cancer that they actually don't. Perhaps he hears enough anecdotes or testimonials, perhaps he is mistrustful of medical science and erroneously concludes that alternative medicine must therefore work, or whatever his thought process is; but an honest conclusion that the null hypothesis has been proven false is a classic Type I error.

TYPE II ERROR: FALSE NEGATIVE

Cynics are those who are most often guilty of the Type II error, the acceptance of the null hypothesis when it turns out to actually be false — it turns out that something is there, but you conclude that there isn't. If you actually do have psychic powers but I am satisfied that you do not, I commit a Type II error. The villagers of the boy who cried "Wolf!" commit a Type II error when they ignore his warning, thinking it false, and lose their sheep to the wolf. The protohuman who hears a rustling in the grass and assumes it's just the wind commits a Type II error when the panther springs out and eats him.

Perhaps somewhere there is a house that actually is haunted, and maybe the TV ghost hunters find it. If I laugh at their silly program and dismiss the ghost, I commit a Type II error. If it were to transpire that the government actually is implementing plans to exterminate millions of citizens in prison camps, then everyone who has not been particularly concerned about this (myself included) has made a Type II error. The invalid dismissal of vitamin megadosing would also be a Type II error if it turned out to indeed cure cancer, or whatever the hypothesis was.

Type I and II errors are not limited to whether we believe in some pseudoscience; they're even more applicable in daily life, in business decisions and research. If I have a bunch of *Skeptoid* T-shirts printed to sell at a conference, I make a Type I error by assuming that people are going to buy, and it turns out that nobody does. The salesman makes a Type II error when he decides that no customers are likely to buy today, so he goes home early, when in fact it turns out that one guy had his checkbook in hand.

Both Type I and II errors can be subtle and complex, but in practice, the Type I error can be thought of as excess idealism, accepting too many new ideas; and the Type II error as excess cynicism, rejecting too many new ideas.

Before talking about Type III and IV errors, it should be noted that these are not universally accepted. Types I and II have been standard for nearly a century, but various people have extended the series in various directions since then; so there is no real convention for what Types III and IV are. However the definitions I'm going to give are probably the most common, and they work very well for the purpose of skeptical analysis.

Type III Error: Answering the Wrong Question

Types III and IV are a little more complicated, but they're just as common as just as important to understand. A Type III error is when you answer the wrong question; and how this usually comes around is when you base some assumption upon a faulty or unproven premise, and so you jump one step ahead and solve a problem that isn't yet the question at hand.

The ghost hunters in the haunted house make a Type III error when they start with the assumption that a ghost makes a cold spot in the room, and so they walk around the haunted house with all sorts of fancy thermometers and collect detailed temperature readings throughout the building. This is great; they've done fine work, and documented it all very nicely, and they correctly reported temperatures. However it is a Type III error, because the question of temperatures has not yet been shown to be relevant, since it has never been established that ghosts affect temperatures.

The conspiracy theorists commit a Type III error when they publish a detailed list of all the locations they've identified as government prison camps. The question is not yet "Where are these camps?" because they skipped over convincingly answering the precedent question of "Do such camps exist at all?" You can produce lists all day long, but until you first prove that each item on the list is actually what you claim it is, the list is of no value.

The vitamin salesman commits a Type III error every time he answers a customer's question about what vitamin is best to take to treat or prevent cancer. He'll no doubt give some such answer and recommend a particular supplement, and perhaps recommend a dosage. This is a Type III error because he's ignoring and skipping the precedent question, which is whether the vitamin in question will treat or prevent the particular cancer in question at all.

TYPE IV ERROR: ASKING THE WRONG QUESTION

While the Type III error is usually committed innocently and with good intentions, the Type IV error — asking the wrong question — often suggests a deliberate deception. By selecting the wrong question to investigate, it's possible to have greater control over the results. Selecting the wrong question is a great way of diverting attention away from the right question.

The producers of ghost hunting TV shows know that they need to produce a program that yields positive results. They also know that they're not going to happen to run into any ghosts or catch anything unexpected on camera. So instead, they frame their program around asking the wrong questions: Can we get interesting readings on our electrical and temperature meters? By structuring their show around the wrong questions, they commit a deliberate Type IV error in order to produce the desired answers.

Conspiracy theorists of all flavors love the Type IV error, as it is one of the most effective tools to build arguments in support of nonexistent phenomena. If the conspiracy theorist wants to convince us that the government is building prison camps to enslave American citizens, it's not necessary to actually ask that question. Instead, ask a whole assortment of related questions that are guaranteed to have positive answers. Are there examples of government corruption? Has the government imprisoned people in the past? Are there laws that permit the government broader powers during times of emergencies? Are there plots of land for which there is no obvious purpose? These questions are all great Type IV errors for the conspiracy theorist.

Similarly, alternative medicine proponents can ask Type IV error questions to suggest that their central claims, which are unevidenced, are actually true. Are there examples of corruption in Big Pharma? Do any natural compounds have therapeutic value? Do scientists rely on grant money? Is medical science big business? Again, these questions are easily answered positively and appear to justify the use of vitamins to treat cancer; when in fact, none of them have any direct relevance to that.

And so there we have it. Four types of reasoning errors, four cases you've heard a thousand times, and will hear a thousand more times. Listen to a few sales pitches, watch a few documentaries on the pseudoscience TV channels, and see if you can spot them. Chances are you will. And, if you can develop enough familiarity with them to spot them when you hear them, you're a leg up on avoiding making these same errors yourself. We all do it, and the better we understand the errors, the better prepared we are to minimize our own such failings.

REFERENCES & FURTHER READING

Kaiser, H. "Directional Statistical Decisions." *Psychological Review.* 1 May 1960, Volume 67, Number 3: 160-167.

Kimball, A. "Errors of the Third Kind in Statistical Consulting." *Journal of the American Statistical Association.* 1 Jun. 1957, Volume 52, Number 278: 133-142.

Mitroff, I., Featheringham, T. "On Systemic Problem Solving and the Error of the Third Kind." *Behavioral Science.* 1 Nov. 1974, Volume 19, Number 6: 383-393.

Mitroff, I., Silvers, A. *Dirty Rotten Strategies: How We Trick Ourselves and Others into Solving the Wrong Problems Precisely.* Stanford: Stanford Business Books, 2009.

Neyman, J., Pearson, E. "On the Use and Interpretation of Certain Test Criteria for Purposes of Statistical Inference: Part I." *Biometrika.* 1 Jul. 1928, Volume 20A, Numbers 1-2: 175-240.

Shermer, M. *The Skeptic Encyclopedia of Pseudoscience.* Santa Barbara: ABC-CLIO, 2002.

41. STAR JELLY

Jellylike blobs have been reported to fall from the sky during meteor showers. What are they?

Cumbria, 2011; Washington, 1994; Massachusetts, 1983; Pennsylvania, 1950; in fact, virtually everywhere on the planet, nearly every year, going back as far as written histories go — star jelly has been reported falling from the skies. It's also been called *star rot* or *fallen star* in a variety of languages, and old texts have referred to it as "a fatty substance emitted from the Earth" and "a mucilaginous substance lying upon the earth". Star jelly is sticky, slimy goo, in puddles or patches, that's said to fall from the skies, usually associated with meteor storms or with shooting stars. It often doesn't last long, drying up and blowing away before it can be properly analyzed.

Star jelly's correlation with falling stars is not universally observed, but many meteors fall without being noticed. One fairly rational Earthly explanation that's been suggested is that star jelly could be chemical waste dumped from the toilets of airliners flying overhead. In more recent years, even conspiracy theorists have gotten in on star jelly, postulating that it is residue from government chemtrail spraying operations. In some cases, witnesses have even reported widespread sickness in towns where the jelly has been found, prompting some to speculate that it may contain some sort of alien virus, thus making it an example of panspermia.

Biologists and other scientists, however, have a wide assortment of less spectacular explanations for the goo. Five minutes on Google reveals that samples, when examined, usually turn out to be some sort of biological growth. But sometimes they don't; it's equally easy to find examples where either no such testing was done, or where a microscopic test found no familiar cells. One thing that's certain is that there is no single substance that accounts for all the many blobs found. Photographs and descriptions are all over the map. Different colors, different textures, different parts of the world. We have to abandon the idea that there is one explanation for star jelly. The various blobs found probably represent many different things.

Many star jelly blobs are probably slime molds. Slime molds are strange organisms, neither fungus nor bacteria, that reproduce via spores. They live in dead plant matter, feeding on the microorganisms found there. The life cycle of slime molds begins as a single cell, which reproduces quite quickly, moving visibly, and can range in size from smaller than a coin to over a meter. During this growth phase, they are wet and slimy, and called a plasmodium. They look a great deal like many star jelly photographs, ranging in color from clear to white to yellow to pink to red. In fact, one common species of slime mold, *Enteridium lycoperdon,* is called *caca de luna* in Mexico, basically "Moon poop". Later in the life cycle, the plasmodium dries up, hardens, splits opens, and releases brown spores, each of which blows away to begin the life cycle anew.

Another highly probably candidate for a lot of star jelly findings is a genus of cyanobacteria (commonly but incorrectly referred to as blue-green algae) called Nostoc, of which there are many species. Nostocs live all over the world as tiny, barely visible colonies of bacteria; but when they get wet, for example in a rainstorm, they swell up to great size and become great gooey lumps or puddles, ranging from clear to yellow to green. Some species are collected and eaten throughout Asia and Central and South America. Notably, Nostoc's common names include *star jelly* and *fallen star.*

There's also a type of slime bacteria called myxobacteria, which travel in packs of cells and actively move through soil. In starvation conditions, intercellular communications prompt the slime bacteria to group together into what's called a fruiting body which can then emerge from the soil. However these jellylike fruiting bodies are tiny, perhaps a millimeter or two in size, and are not a good explanation for most star jellies.

Some fungi look superficially like a few of the star jellies reported, such as *Phlebiopsis gigantea*. Although this and other fungi can often appear and grow quite quickly, and can do so simultaneously scattered over a wide area, many star jelly enthusiasts point out that fungi are neither jellylike nor sticky. Fungi are generally dry, with a consistency more like bread than like slime. However many fungi do secrete a smelly slime, generally thought to attract flies that help spread the spores. One whole group of mushroom are in fact called slime caps. But after comparing many photographs of slime-producing fungi with star jelly specimens, I find it very hard to accept that one could be mistaken for the other.

But there is another interesting candidate, once thought to be related to fungi but now considered to be separate. Bryozoa are a whole phylum of creatures many people have never heard of. There are about 4,000 species, and they exist in colonies of interdependent individuals. We call such colony-dependent individuals zooids. Most of these zooids are tiny but not microscopic, about a half a millimeter long, and they secrete exoskeletons. In some species, this is relatively solid, giving the colony a coral-like or plant-like appearance; in others, this is gelatinous, turning the whole colony into a wet, sticky blob of nonliving gel with a living surface of zooids.

Bryozoan blobs live in the water, some in salt, some in fresh; and attach themselves to underwater objects. Some species move up to several centimeters a day. The largest colonies have been observed to be a meter across. One such blob famously surfaced in a lake in Newport News, VA in 2010, prompting all sorts of speculation from people who had no idea what the heck it might possibly be.

Other racier natural substances include unfertilized frog spawn or even deer sperm, both of which do appear clumped on the ground in nature and look very similar to a lot of the photos of star jelly found on the Internet. As surprising and unlikely as they may sound, it's more than likely that at least some of the many sightings throughout history were these simple substances.

The suggestion that some star jelly is actually human waste dumped from airplane toilets is a poor fit. First of all, airplanes don't now (and never have) discharged toilet waste by dumping it in mid-air; however, it is possible for their systems to leak. On the few occasions when this has happened, the blue liquid forms ice on the fuselage, which can break off and fall, usually when the plane is descending into warmer air before landing. There have been about a half dozen reported cases in the past few decades of such chunks actually damaging roofs. But these ice chunks are not gelatinous, and their blue color is not a good match for many of the reports of star jelly.

The anecdotal reports of mass illness striking residents after falls of star jelly also fail to stand up to scrutiny. It's fine to say that lots of people in town got sick, and we can even grant that they did; but what were they sick with? What ties the sickness to the star jelly? Virtually every town has some virus being passed around at any given time; and no matter what event happens, it's almost always possible to correlate it with a current breakout of some sickness. It certainly is possible to connect an epidemic with a cause, and so far, this has not been done in a single star jelly case. It's been pure anecdotal speculation; there's never once been a diagnosis of a pathogen tied to star jelly. Especially given that someone in any town is always sick, there has not yet been any compelling science-based reason to conclude that star jelly produces ill effects.

Perhaps the most important weak assumption about star jelly is its origin: having fallen from the sky. News reports almost always state the substance rained out of the sky, and indeed, most people talking to the reporters honestly believe that it did. Most of them are probably assuming it, but others have a firmer conviction. They saw the ground with no star jelly, then maybe regular rain fell or maybe nothing fell, and then they observed the star jelly. No other *apparent* explanation was possible, so the notion of its having fallen is often taken for granted as an observation. But the fact is that there is *no* testable, non-anecdotal evidence that any star jelly has *ever* fallen from the sky.

Exactly as we found when *Skeptoid* studied the stories of frogs and fish raining out of the sky, the observation of falling seems so obvious that eyewitnesses firmly believe that it happened and tend to report it, whether they actually saw it or not. Verbal reports have included fish flipping about in rain gutters on rooftops, despite no corroborating evidence. So it is not at all surprising — in fact it's *expected* — that some star jelly witnesses will report actually having seen it fall from the sky. Maybe it did, but science does not allow us to assume that all verbal reports precisely and inerrantly describe what actually took place. We have to rely on what we can test and what we can know; and so far, star jelly is not *known* to fall from the sky at all.

But what about DNA testing, shouldn't that tell us for sure whether star jelly is an Earthly organism or something else? There's a good reason that DNA tests are rarely done on star jelly. Unlike their television portrayal, DNA tests — especially in past years when famous star jelly events have happened — have been prohibitively expensive and involved. Star jelly would have to be taken to a university equipped to do such testing, and would require lots of time and money. Who is supposed to pay for that? Volunteers stepping forth with their checkbooks open have been few. Star jelly cases have rarely excited the interest of scientists sufficiently to get them to front the costs from their own pockets, simply because there are so many possible common and unremarkable explanations.

My summation on star jelly is that it's a mistake to consider it a single phenomenon. Every case of some jellylike slime found on the ground is unique, and should be treated as such. There are so many possibilities; every given case, on visual and tactile data alone, offers at least two or three possibilities from Earth's own taxonomic kingdoms. Is any star jelly actually from the stars? Maybe some of it is, but it would be the exception not the rule; and has yet to be supported with evidence.

REFERENCES & FURTHER READING

Adams, C. "Did Mrs. Sybil Christian of Frisco, Texas, find blobs from space on her lawn?" *The Straight Dope.* Creative Loafing Media, Inc., 31 May 1985. Web. 22 Feb. 2012. <http://www.straightdope.com/columns/read/521/did-mrs-sybil-christian-of-frisco-texas-find-blobs-from-space-on-her-lawn>

Editors. "Lake District Invaded by Alien Jelly Last Seen in Scotland." *The Daily Record.* DailyRecord.co.uk, 22 Oct. 2011. Web. 22 Feb. 2012. <http://www.dailyrecord.co.uk/news/weird-news/2011/10/22/lake-district-invaded-by-alien-jelly-last-seen-in-scotland-86908-23506580/>

Editors. "Star Jelly – Goopy Death from Beyond Space." *The Skeptical Viewer Forums.* SkepticalViewer.com, 5 Jun. 2010. Web. 25 Feb. 2012. <http://www.skepticalviewer.com/forums/possibly-paranormal/star-jelly-goopy-death-from-beyond-space/>

Marshall, M. "Slime Time." *The Last Word.* New Scientist, 7 May 2008. Web. 23 Feb. 2012. <http://www.newscientist.com/blog/lastword/2007/10/slime-time.html>

Ostry, M., Anderson, N., O'Brien, J. *Field Guide to Common Macrofungi in Eastern Forests and Their Ecosystem Functions.* Newton Square: United States Forest Service, 2010.

Radford, B. "Mysterious Lake Blob Identified as Alien Bryozoan." *Live Science.* TechMediaNetwork.com, 8 Nov. 2010. Web. 22 Feb. 2012. <http://www.livescience.com/11086-mysterious-lake-blob-identified-alien-bryozoan.html>

42. THE BEALE CIPHERS

Treasure hunters comb Virginia search for a legendary hoard of gold and silver.

If you see someone digging with a shovel under the moonlight in Bedford County, Virginia, chances are you've come across a treasure hunter. For more than a century, hopeful treasure seekers have combed these green hills, searching in vain for a fantastic treasure said to have been buried here, as described in a mysterious coded document. The story goes that a man named Thomas Beale discovered a fabulous wealth of gold and silver in 1818 in what is now Colorado, and along with his company of thirty partners, brought it back to the east and buried it in Virginia. Beale wrote three encoded letters: one giving the exact location of the treasure, a second giving its detailed description, and a third giving the names and contact information of the thirty partners. Only one of the letters — the second describing more than four tons of gold, silver, and jewels — was ever deciphered. It has tempted the greedy and adventurous ever since.

The world first learned of the Beale ciphers in 1885 with the limited publication of a pamphlet in Lynchburg, Virginia entitled *The Beale Papers: Containing Authenticated Statements Regarding the Treasure Buried in 1819 and 1821, Near Bufords, in Bedford County, Virginia, and Which Has Never Been Recovered.* It was written by James B. Ward. A 1981 article in *Smithsonian* described him as a "gentleman of independent means". In *The Beale Papers,* Ward provided the three ciphers and told the story of how he came into possession of them. But after many years of toil and neglect of family and responsibility, he was:

...compelled, however unwillingly, to relinquish to others the elucidation of the Beale papers, not doubting that of the many who will give the subject attention, some one, through fortune or accident, will speedily solve their mystery and secure the prize which has eluded him.

In Ward's account, the man Thomas Jefferson Beale had been a regular customer at the Washington Hotel in Lynchburg, kept by a man named Robert Morriss. Morriss died in 1865, twenty years before Ward revealed all his secrets in his pamphlet. In 1822, Beale returned to Morriss' hotel and gave Morriss a locked iron strongbox for safekeeping. Morriss gave the box little thought until he received a letter from Beale in May of that same year. The letter advised him of the great importance of the box's contents, and added:

> *Should none of us ever return you will please preserve carefully the box for the period of ten years from the date of this letter, and if I, or no one with authority from me during that time demands its restoration, you will open it, which can be done by removing the lock. You will find, in addition to the papers addressed to you, other papers which will be unintelligible without the aid of a key to assist you. Such a key I have left in the hands of a friend in this place, sealed, addressed to yourself, and endorsed not to be delivered until June, 1832.*

Morriss waited the ten years but never saw or heard from Beale again. Neither did the promised key ever arrive. So in 1845, Morriss finally opened the box on his own, 23 years after receiving it. He found the three encoded letters, plus another letter addressed to himself in which Beale gave his account of finding the treasure. Beale's company of thirty adventurous partners, all Virginians, had been hunting north of Santa Fe, when they happened across a ravine full of gold. No specific description of the gold was given, so we're left to guess whether it was bags of coins, raw nuggets, or natural ore. Beale states that it took eighteen months to extract it, so it seems probable that it was naturally occurring. The partners contrived to share in the wealth equally, but to first transport it cross country and hide it in Virginia, where it would be more secure than in the relatively lawless territories. Beale and nine companions took the first load to Virginia and buried it, then returned to the other twenty who still toiled. All thirty then took the final load and buried it along with the rest.

Morriss was never able to decipher the papers; and so in 1862, he turned the whole affair over to Ward, with the understanding that if Ward ever recovered anything that they should share in it. It was well after Morriss' death that Ward finally decoded the second paper. The key to the cipher was discovered to be the United States' Declaration of Independence; and the cipher solved by taking each number from the code, counting that many words in the Declaration, and using the first letter of that word. The letter said:

> I have deposited in the county of Bedford, about four miles from Buford's, in an excavation or vault, six feet below the surface of the ground, the following articles, belonging jointly to the parties whose names are given in number "3," herewith:

> The first deposit consisted of one thousand and fourteen pounds of gold, and three thousand eight hundred and twelve pounds of silver, deposited November, 1819. The second was made December, 1821, and consisted of nineteen hundred and seven pounds of gold, and twelve hundred and eighty-eight pounds of silver; also jewels, obtained in St. Louis in exchange for silver to save transportation, and valued at $13,000.

> The above is securely packed in iron pots, with iron covers. The vault is roughly lined with stone, and the vessels rest on solid stone, and are covered with others. Paper number "1" describes the exact locality of the vault so that no difficulty will be had in finding it.

And that was as far as Ward ever got before publishing the papers. Bedford County has been dug up regularly ever since, but no treasure was ever found. Many researchers have fact checked and studied *The Beale Papers,* none more thoroughly than Joe Nickell, as described in his article in a 1982 issue of *The Virginia Magazine of History and Biography.* What we've learned is that Beale and his treasure are almost certainly a product of James Ward's imaginative fiction.

For one thing, Nickell found no record of a Thomas Jefferson Beale from Virginia, at least not of the right age. For another, there were impossibilities in the dates in *The Beale Papers;* such as Beale's having first been a customer of Morriss' at the hotel in 1820, and written a letter to him at the hotel dated 1822, when Morriss did not become the hotel's proprietor until 1823; and Beale's use of the words *stampede* and *improvise,* neither of which existed in the English language until much later.

Indeed, the name Thomas Jefferson Beale suggests an inside joke. Thomas Jefferson was, of course, the author of the Declaration of Independence, the very document Ward claimed to have stumbled upon "by accident" (in his words) as the key to the cipher. The name Beale suggests Edward Fitzgerald Beale, who became famous when he crossed what was then Mexican territory in disguise to transport the first samples of California gold from the west to the east, 37 years before Ward's book.

How likely is it that neither Beale nor any of his thirty companions would have wanted to spend and enjoy their fabulous wealth, but were instead content to leave it buried for the rest of all their lives? Ward wrote that he suspected they'd been killed by Indians; ridiculously proposing that all thirty would have walked away from their fortune to go camp out in the territories.

Compare the Beale report to what happened when gold actually was discovered in California thirty years later in 1848.

Word spread like mad, compelling hundreds of thousands of men to storm west in one of the most important events in American history, the California gold rush. Yet Beale and his men told no one, made two cross-country trips trucking four tons of gold and silver across the nation with nobody noticing, and even traded large amounts of silver for jewels in St. Louis without piquing any interest or inquiry. It strains credibility.

Mineralogy also casts doubt on the story. Gold and silver ore do not look anything like gold and silver, and would not be recognizable to any but experienced prospectors. Once mined, the ore requires lengthy and substantial processing to reduce it to just the precious metals. Beale's account had the men simply recognizing gold and picking it up, for eighteen months. This means that it would have had to be a placer mine, where small nuggets appear naturally in the sediment. This strike would have made it the single richest placer district in Colorado's mining history, meaning these thirty men out-mined all of those tens of thousands who followed in Colorado's later silver and gold rushes. Further, gold and silver placer pure enough to look distinct from one another are never found in the same place. If it is, the metals would most likely be alloyed, and not recognizable separately as gold and silver.

The length of the third cipher, 618 characters, tells us that Beale's statement that it contains the names, addresses, next of kin, and their contact information for thirty men is unlikely. That's a lot of information to squeeze into twenty characters for each man. From a practical standpoint, it's impossible; so at the very least, either Ward or Beale was lying or hopelessly incompetent at cryptography. The former is probably true; cryptographers studying the case have been unanimous that for Ward to have found such an obscure solution to the second cipher is not plausible. Combined with several errors in the coding and decoding that Ward did not appear to catch, the only reasonable explanation is that the person who solved the cipher also created it at the same time. Beyond any reasonable doubt, James

Ward was the creator of the ciphers, not Morriss, and not the apocryphal Beale.

Nickell and later researchers, notably Louis Kruh in the journal *Cryptologia,* also found compelling stylometric evidence that the authors of Ward's pamphlets and Thomas Beale's letters were one and the same person, especially when compared against three contemporary control authors. We can only conclude that Ward wrote both the letters and the ciphers he attributed to Beale, and included Robert Morriss only to anchor his story to a grain of truth, and because Morriss had been dead for twenty years and was unlikely to dispute the account.

After its publication, Ward tried to downplay the tale, claiming that all remaining copies of the pamphlet had been destroyed in a print shop fire, despite researchers finding no newspaper records of any such fire. Only the first few pamphlets ever got out, and once they did, it appears that Ward realized he'd created a monster with a greater effect than he'd anticipated. Ward had been friends with the Buford and Morriss families, and it's perhaps most probable that the unexpected attention changed his mind about promoting his fictional story at the expense of his friends. It's most conspicuous that he never reprinted the pamphlet, despite great demand, great potential for sales, and availability of five other print shops in Lynchburg.

It's hardly surprising that none of the Bedford County treasure hunters have yet turned up any evidence of Beale's fabulous hoard. It may never have been more than a small joke that took an unwanted direction, but it's become a firm fixture in the annals of American legend.

REFERENCES & FURTHER READING

Anonymous. "Pursuit." *Beale Ciphers Analyses.* Anonymous, 1 Jan. 1986. Web. 5 Mar. 2012.
<http://www.angelfire.com/pro/bealeciphers/Graphics/Pursuit.pdf>

Anonymous. "The Mysterious Treasure of Thomas Beale." *h2g2 Guide Entries*. British Broadcasting Corporation, 18 Jul. 2011. Web. 6 Mar. 2012. <http://news.bbc.co.uk/dna/place-lancashire/plain/A83646156>

Daniloff, R. "A Cipher's the Key to the Treasure in Them Thar Hills." *Smithsonian*. 1 Apr. 1981, Volume 12, Number 1: 126-144.

Gillogly, J. "The Beale Cipher: A Dissenting Opinion." *Cryptologia*. 1 Apr. 1980, Volume 4, Number 2.

Nickell, J. "Discovered: The Secret of Beale's Treasure." *The Virginia Magazine of History and Biography*. 1 Jul. 1982, Volume 90, Number 3: 310-324.

Viemeister, P. *Beale Treasure: A History of a Mystery*. Bedford: Hamilton's, 1987.

Ward, J. *The Beale Papers: Containing Authenticated Statements Regarding the Treasure Buried in 1819 and 1821, Near Bufords, in Bedford County, Virginia, and Which Has Never Been Recovered.* Lynchburg: Virginian Book and Job Print, 1865.

43. DE LOYS' APE

*A geologist claimed to have discovered a new species of
great ape in Venezuela in the early 20th century.*

Today we're going to point the skeptical eye at one of the
many cryptids known from some scant evidence. Our subject is
De Loys' Ape, also called Ameranthropoides loysi. It is said to
be an unknown species of primate, walking bipedally through
the jungles of South America like a hominid, having no tail like
an ape, but bearing the physical appearance of a monkey. It
stands one and a half meters tall (four and a half feet) and is
known primarily from a single photograph, taken by an oil expedition in Venezuela sometime around 1920.

The world had known that Venezuela held vast oil reserves
since the first Spanish conquerors arrived in the 16th century,
but it wasn't until the early 1900's that the first modern oil
wells were drilled. Following World War I, oil prospecting exploded in Venezuela, such that by 1929 it was the world's largest oil exporting nation. In 1917, Swiss geologist François de
Loys was only one of countless Europeans brought in by the oil
companies to prospect. De Loys spent the better part of four
years in Venezuela, traveling about and prospecting. It was during this period that something happened that would make de
Loys famous forever.

If you do an image search for de Loys' Ape or Ameranthropoides loysi, you'll find plenty of results. The creature is
known by a single old black and white photograph showing a
dead specimen, propped up with a stick under its chin, posed in
a sitting position on a wooden crate. The face is ugly; quite ee-
rily so, in fact. Physically, the animal looks a lot like a spider

monkey, but is said to have been substantially larger than any known spider monkey. According to the story, de Loys and his prospecting party were in the jungle. De Loys himself told the tale in a 1929 issue of the *Illustrated London News*, which was reprinted in *The Washington Post* and titled "Found at Last - The First American":

De Loys' ape

...The jungle swished open, and a huge, dark, hairy body appeared out of the undergrowth, standing up clumsily, shaking with rage, grunting and roaring and panting as he came out onto us at the edge of the clearing. The sight was terrifying...

The beast jumped about in a frenzy, shrieking loudly and beating frantically his hairy chest with his own fists; then he wrenched off at one snap a limb of a tree and, wielding it as a man would a bludgeon, murderously made for me. I had to shoot.

My Winchester got the best of the situation. Riddled with bullets, the great body soon fell on the ground almost at my feet, and quivered for a while. He gathered his arms over his head as if to hide his face and, without a further groan, expired.

De Loys examined the creature and discovered that it had no tail, which is characteristic of great apes and not of monkeys. He found that it had 32 teeth, which also differs from New World monkeys who have 36. From there, most versions of the story state that de Loys and the survivors of his party tried to preserve the head and hide of the beast, but that both were lost or abandoned during the group's subsequent misadventures and clashes with natives.

Knowledge of de Loys' ape story might never have reached the public consciousness were it not for the involvement of the French anthropologist Georges Montandon. Montandon's interest was race, in fact he later developed a racial taxonomy to classify and categorize many different races of humans. I was not able to find how and when de Loys and Montandon first met, but 1929 is when he first saw de Loys' photograph and heard his tale. As de Loys told the story to the public in the newspapers, Montandon told the story to the scientific community. He published an article in the *Journal de la Société des Américanistes,* and presented a talk to the French Academy of Sciences. Montandon declared that de Loys' ape was a new species, and gave it the name Ameranthropoides loysi.

Ultimately, Montandon failed to convince scientists, and Ameranthropoides loysi never made it into any taxonomy other than his own. But the more visible campaign, the public campaign waged by de Loys himself, was quite successful. De Loys' ape has been a pinnacle of cryptozoology ever since, and is featured in virtually every book on the subject.

Coming into this topic, the question I wanted to answer was how we should go about determining the truth of de Loys' ape. As with most questions like this, it's quite possible that we can never know the answer for a certainty, but we can at least get a pretty good idea. The three most likely possibilities are that it truly is a new species; that it was deliberately hoaxed somewhere along the line; or the third and most common possibility, that it was an honest misidentification combined with trumped-up reporting. This is the null hypothesis: that nothing extraordinary is proved to have happened, either a deliberate hoax or an actual new discovery.

By digging deeper into the histories of François de Loys and Georges Montandon, we can hope to find some additional clues into their motivations. Montandon appears to have been an ordinary anthropologist, and de Loys an ordinary geologist; for both men, the de Loys' ape episode is the only extraordinary event in their histories. But when it happened, both seemed to go to extraordinary (and unscientific) lengths to prove the creature was a new discovery. In fact, in de Loys' newspaper report, he even stepped far enough outside his geology background to assert this:

> Until my discovery of the American anthropoid, we could only imagine that man migrated to these shores. But now, in the light of this discovery, it is obvious that the failure of the otherwise well established principle of evolution when it was applied to America was due only to imperfect knowledge. The gap observed in America between monkey and man has been eliminated; the discovery of the Ameranthropoid has filled it.

Together with Montandon's aggressive promotion of the discovery in the scientific community, it does appear that both men went out of their way to prove something which wasn't well supported.

We also want to look at the actual primate population in South America to see if something like de Loys' ape might exist. So far, there are no other reports of anything like it. Some cryptozoologists have looked for local legends that might serve as a candidate match for de Loys' ape. One has been consistently mentioned, the *didi*. Most didi stories are reminiscent of Bigfoot; a large, hairy man-beast, often fierce or aggressive. There are virtually zero bibliographic references to the didi in the English language outside of the cryptozoology literature. It's certainly not a creature for which any empirical evidence exists, and thus it would be premature to characterize it as a possible explanation for de Loys' ape. In order to suggest that creature X matches creature Y, we have to know the characteristics of each. Didi stories differ widely from one another, so it's not logical to say that it has any established characteristics.

Another way we can evaluate de Loys' story is to look at all its details, and compare them with known history. Does his story pass fact checking? Does it fit well into the context of what was happening in Venezuela around 1920? The story does come under suspicion. Notably, in his newspaper article, he claimed that no fewer than seventeen of his men had been killed by the arrows of native tribesmen, and that he himself had been shot in the thigh, and was still lame from this wound at the time. The place where he says he was attacked by the creatures is only about 15 kilometers west of La Fría, a city founded in 1853, capitol of the municipality in which oil was discovered. La Fría and other towns in the area are where the oil companies based their expeditions. In short, de Loys' location simply wasn't as remote as most modern retellers indicate. It's often said that he spent four years in the jungle, as if it was a single long expedition, but it wasn't. He spent four years making much shorter prospecting trips *into* the jungle.

How likely is it that a geology party would allow seventeen men to die without simply making the short return trip to La Fría or some other town? How likely is it that de Loys would have continued prospecting if *any* men had actually been violently killed on the job? The job simply wasn't worth men's lives, and at no time was de Loys in so remote a position that he could not easily have returned to civilization. In short, his story, as printed, is almost certainly a gross exaggeration in the style of the popular adventure fiction of the day.

There are also conflicting accounts of the famous photograph's history. In 1962, a Dr. Enrique Tejera read an account of de Loys' ape in a magazine called *The Universal*, and wrote to its author. The letter was published. Tejera said that he'd been in Venezuela with de Loys, working for the oil companies as a doctor, and the monkey photograph had a very different history. According to him, de Loys had been something of a practical joker, and had been given a local spider monkey called a Marimonda (white-fronted spider monkey, *Ateles belzebuth*), whose tail had become infected and been cut off. While they were in the city of Mene Grande in 1919, the monkey died, and that's when de Loys propped it up on the crate and took the photograph.

Of course we don't have any way to know which of these two stories is true (or even whether both are false), but the Tejera version is certainly more consistent with the photo. The monkey is sitting on a crate which would have to be very large if the monkey was as tall as de Loys said, and it's sitting in a clearing on a riverbank, among the stumps of plants that have obviously been cut away. Yet in de Loys' story, he and his men had only stopped there briefly to rinse off from "the day's struggle across the jungle", hardly consistent with well-prepared clearings and large heavy crates.

In conclusion, I'm satisfied that there's enough wrong with de Loys' story to assert that it was, at least in large part, fabricated; but I wouldn't say that this can be proved. Regarding Montandon's involvement, my own assessment is that he may or may not have been deliberately deceptive. The story he promoted was in line with his normal research, and he worked in a science that was in its adolescence if not its infancy; but he was willing to put his professional reputation on the line. I would probably camp in the jungles of Venezuela and not be worried about marauding ape-men; but I might just tip my turn-of-the-century flask in applause of François de Loys and his pet spider monkey with the infected tail.

REFERENCES & FURTHER READING

De Loys, F. "Found at Last - The First American." *The Washington Post.* 24 Nov. 1929, Newspaper: SM14.

Heuvelmans, B. *On the Track of Unknown Animals.* New York: Hill and Wang, 1959.

Lieuwen, E. *Petroleum in Venezuela: A History.* Berkeley: University of California Press, 1954. 18-32.

Montandon, G. *La Race Les Races.* Paris: Payot, 1933.

Silverberg, R. *Scientists and Scoundrels: A Book of Hoaxes.* New York: Crowell, 1965. 178-187.

Urbani, B., Viloria, A. *Ameranthropoides loysi, Montandon 1929: the History of a Primatological Fraud.* Buenos Aires: Libros en Red, 2008.

44. Are Vinyl Recordings Better than Digital?

Many audio aficionados split into two camps, those supporting modern digital audio, and those supporting vinyl records.

For as long as there have been competing standards — horses versus steam, paper versus parchment, Android versus iPhones, Whigs versus Tories — fanatics have taken sides and promoted them as superior with nearly religious passion. The comparison of sound quality between vinyl records and digital recordings stands tall among these platform debates. Nearly all audio enthusiasts take one side or the other, some openly and with zeal, most with subtlety and qualifying their preference through acknowledgements of the pros and cons of each. Either way, one basic question supersedes either preference: Does it make any detectable difference?

Again, those full of zeal, on both sides, assert that the difference is detectable, implying that they would be able to tell. In a few moments we'll take a look at some of the testing that has been done to study this claim. But first, a quick overview of the salient technical points.

The principal difference is the nature of the storage medium, which is either analog or digital; a smooth-flowing waveform as cut into the grooves of a vinyl record, or digital representation of the recorded sound with numeric amplitudes sampled at a high frequency. There's an exquisite elegance to the way that a stereo signal — two discreet, simultaneous channels of music — can be encoded into a single groove that one needle follows. As the groove moves side to side, a single channel is produced, with its frequency determined by the speed at which the needle is pushed left and right, and its amplitude determined by how far it's pushed. To add a second channel, we bring in a second axis of movement: vertical in addition to horizontal. Tip them both over at 45°, and we have a groove that varies in depth as well as in its horizontal axis. How fast and how far the needle vibrates down to the left describes the signal in the left channel; how fast and how far the needle vibrates down to the right describes the signal in the right channel. Adding the two signals together produces the instruc-

tions for how the groove is to be cut; at every instant, there is one smoothly flowing waveform describing the left channel, and a second describing the right channel. It's a beautiful system.

A digital audio recording is defined by two basic parameters: the sample rate, which is how many times per second the height of the waveform is sampled; and the resolution, which is the number of possible levels that can be measured at each sample. For a compact disc, this resolution is 16-bit, when means that the height of the waveform is measured, at each step, on a scale of 0 to 65,535, which is very precise. This measurement is performed at a sample rate of 44,100 times per second. This number is chosen because it's just over twice the highest frequency that the best human ears can hear, which is around 20,000 Hertz. A formula called the Nyquist rate shows that this is the minimum sample rate needed to produce the full range of human hearing. Lay each word of sixteen zeros or ones end to end, double it because there is a separate measurement for each stereo channel, stream them past at 44,100 measurements per second, and the speed at which those bits go by is called the bit rate. The higher your bit rate, the higher resolution and sample rate can be used. If this stream is to be recorded on a compact disc, it goes through another conversion to change it into a completely different series of ones and zeros that can be more accurately read by the laser. If it's stored on a computer, it can be algorithmically compressed via any of a number of different schemes, producing tradeoffs between file size and preservation of data.

So there our battle lines are drawn. There are myriad things one could say in addition to each. Vinyl and digital both have good points and bad points. But here's the reason why the entire debate is stupid: *whether the music is stored on vinyl or a CD is just not that important a part of the overall system.* It's like deciding which of two different cars is best by comparing their spark plug wires. There are many, many variables in the process of playing recorded music that noticeably affect the sound, from the microphones, to the mixing, to the mastering, to the quality of the playback hardware, the amplifier, and (far and away most important) the quality of the speakers and characteristics of the listening room; whether the recording was vinyl or CD is simply *not* one of these important variables, with apologies to the zealots. Both methods are easily far superior to any differences the human ear might hope to distinguish.

A lot of vinyl proponents say that the difference is subjective, for example that it sounds warmer or just better. Digital proponents tend to point out objective difference, such as the fact that a digital signal can accommodate a higher dynamic range, which is the difference in loudness between the quietest and loudest parts of the recording. But can they actually tell the difference under controlled conditions?

Well, unfortunately, this is a bit like asking which race car driver is most talented if you put them into identical cars. That car would always suit one driver's style and preferred setup better than the other. Finding an identical recording on vinyl and on CD to compare doesn't really exist. In the early days of CDs, record companies sometimes didn't bother making new masters of the old recordings; they used the same masters that had been used to press the vinyl. The results were CDs that sounded tinny or thin. The master suited vinyl, not digital. Now mastering engineers will almost always make a new master designed for the intended medium. A master is a special mix designed by an engineer who knows who's going to be listening, how they're going to listen, what other music it needs to sound good against, and so on. The separate instrument tracks might be individually equalized, spread across the stereo spectrum, or have a dozen other parameters applied. Thus, a CD and a vinyl pressing of the exact same recorded performance are likely to be very different. If they're not, that means an inappropriate master was used for one or the other, and the test will be biased.

Moreover, the vinyl playback method includes giveaways: clicks and pops, hissing, and other noise produced by the mechanical playback experience. Indeed, much of what's often lauded about vinyl recordings — such as the "richer, warmer" sound — is not a result of accurate reproduction, so much as it is an artifact of the playback mechanism itself.

It's a hard science fact that digital is capable of reproducing higher frequencies than vinyl, above the range of what most people can hear. But, can people distinguish whether a piece of music contains those frequencies or not? According to research performed at Japan's NHK Laboratories in 2004, the answer seems to be no. They took 36 people and ran 20 tests with each. Only a single 18-year-old girl was able to beat random chance, and so they retested her separately, but the effect disappeared. Nevertheless, the researchers issued a somewhat qualified conclusion that they could "neither confirm nor deny the possibility that some subjects could discriminate between musical sounds with and without very high frequency components." Whether that recording is vinyl or digital, any frequencies it may or may not have above 20,000 Hz make no difference.

Controversy also exists between various digital formats, lending credibility to the whole format war concept. Two high end consumer digital formats, Super Audio CD and DVD Audio (technically Direct Stream Digital and Pulse Code Modulation), have been bantered back and forth by industry experts. But in 2004, a paper presented at the 116th Audio Engineering Society conference in Berlin found that:

> ...No significant differences could be heard between DSD and high-resolution PCM (24-bit / 176.4 kHz) even with the best equipment, under optimal listening conditions, and with test subjects who had varied listening experience and various ways of focusing on what they hear. Consequently it could be proposed that neither of these systems has a scientific basis for claiming audible superiority over the other. This reality should put a halt to the disputation being carried on by the various PR departments concerned.

In 2000, some excellent research was published in the *Bulletin of the Council for Research in Music Education* where subjects listened to digital and analog recordings of the same concert performance, recorded unequalized and unmixed especially for this test. They were able to switch back and forth between the two at will, and everything was blinded and well controlled. Overall, the digital version was preferred in all ten scoring areas. However the recording media for this test were compact disc and cassette tape, so it's not directly comparable to a vinyl record. The researchers concluded:

> Results showed that music major listeners rated the digital versions of live concert recordings higher in quality than corresponding analog versions. Participants gave significantly higher ratings to the digital presentations in bass, treble, and overall quality, as well as separation of the instruments/voices. Higher

rating means for the digital versions were generally consistent across loudspeaker and headphone listening conditions and the four types of performance media.

To summarize the science, digital is the superior reproduction format, but analog (particularly vinyl) offers a particular type of sound that some people prefer. I liken it to a Ferrari versus a Mustang. They may have different metrics, but the people who like them for what they are don't care so much about that.

The best argument in favor of vinyl recordings need not be bolstered by unsupported claims about the technical quality of the recording, and that's the physical, tangible experience. Lowering a needle onto a record engraved with an actual audio waveform is comparable to building your own hot rod with greasy hands and case hardened tools. Its performance compared to that of a factory produced BMW is simply not relevant. It's about an experience, not about metrics or tabulated results. More senses are involved: the smell of the album cover, the touch of lowering the tone arm into the groove, the sight of the stroboscope indicating the precise turntable speed. It's a full experience to which the listener must dedicate focused attention and time. Vinyl records are a hands-on, personal connection to the actual audio, and that's something no amount of digital perfection can replicate. You can debate the validity of that connection all you want, and you'll find that it's a metaphysical, philosophical issue. There is no logic or practical connection. But some things, these types of connections — those for which no practical, quantitative explanation exists — are sometimes the most important.

REFERENCES & FURTHER READING

Blech, D., Yang, M. *Convention Paper 6086: DVD-Audio versus SACD: Perceptual Discrimination of Digital Audio Coding Formats.* Berlin: Audio Engineering Society, 2004.

Gabrielsson, A., Hagerman, B., Bech-Kristensen, T., Lundberg, G. "Perceived sound quality of reproductions with different frequency responses and sound levels." *Journal of the Acoustical Society of America.* 1 Jan. 1990, Volume 88: 1359-1366.

Geringer, J., Dunnigan, P. "Listener Preferences and Perception of Digital versus Analog Live Concert Recordings." *Bulletin of the Council for Research in Music Education.* 1 Jul. 2000, Number 145: 1-13.

Lipshitz, S. "The Digital Challenge: A Report." *Boston Audio Society.* Boston Audio Society, 1 Aug. 1984. Web. 18 Mar. 2012. <http://www.bostonaudiosociety.org/bas_speaker/abx_testing2.htm>

Liversidge, A. "Analog versus Digital: Has Vinyl Been Wrongly Dethroned by the Music Industry?" *Omni.* 1 Jan. 1995, Volume 17, Number 5: 28.

Nishiguchi, T., Hamasaki, K., Iwaki, M., Ando, A. *NHK Laboratories Note No. 486: Perceptual Discrimination between Musical Sounds with and without Very High Frequency Components.* Tokyo: Japan Broadcasting Company, 2004.

Repp, B. "The Aesthetic Quality of a Quantitatively Average Music Performance: Two Preliminary Experiments." *Music Perception.* 1 Jan. 1997, Volume 14, Number 4: 419-444.

Sourisseau, U. "Stereo Disc Recording." *Record Your Own Vinyl Discs.* Souri's Automaten, 11 Apr. 2001. Web. 20 Mar. 2012. <http://www.vinylrecorder.com/stereo.html>

45. CATCHING JACK THE RIPPER

A look at what is and isn't known about history's most infamous serial killer.

Today we're going to pay a visit to London, specifically the Whitechapel district, back in 1888. It was a crowded, filthy, uncomfortable place, still stinging from the deadly Bloody Sunday riots of 1887 when Irish protesters clashed with police. Political unrest, unemployment, crime, rampant disease, and hunger were compounded with racial tensions between the Irish working class and Jewish refugees from Eastern Europe. Amid all this squalor, it's hardly surprising that violence and murder were common. In fact, those days and that place produced history's single most infamous murderer: Jack the Ripper. We're going to take a look at the common knowledge about Jack, and see how it compares to what's actually known and what's popularly believed. Who was he, how many did he kill, and how did we learn what we know?

Poverty was so severe in Whitechapel — at the time, the acknowledged slum of London — that prostitution became one of the leading industries. It was a desperate move by a population, desperate from actual hunger and deprivation, to bring money into the community from London's wealthier West End. This workforce of an estimated 1200 women commuted every day and night on foot through unlit alleys filled with drunken factory workers and as many as 3000 homeless people in this neighborhood alone, and their victimization was all too commonplace. Police grouped eleven murders of prostitutes, committed between 1888 and 1891 and all apparently related in some way, and referred to them as the Whitechapel Murders.

At the time, the presumed killer was known to the press as Leather Apron.

Later, investigators separated five of the Whitechapel Murders away from the rest, believing them to be the work of a single distinct individual. The modus operandi was the same: the victims were first silenced and killed by strangulation, their throats were cut, and their bodies mutilated in similar ways, to varying degrees depending on how private was the location and how much time the killer probably had.

Today, nearly a century and a half later, Jack the Ripper still grips the public's attention. Practically every year, a number of books and documentaries are produced naming some new suspect or claiming a fresh analysis on some piece of evidence. Please don't email me about this new book or that new book and its particular claims and its evident authority; the Library of Congress currently lists at least 181 different books with *Jack the Ripper* in the title; every one I've looked at introduces yet another suspect, and they can't all be right. Many believe he was an aristocrat or educated man from the West End, and some believe he was no more than a local thug. Today's enthusiasts and armchair investigators call themselves Ripperologists.

The Ripper was never seen or heard and left no evidence that survives today, with the possible exception of letters received by investigators and newspapers claiming to be from him, and it's the signature of one of these letters that is the first recorded appearance of the nickname Jack the Ripper. But at the time, eyewitness reports abounded. Nine days after the Ripper's victim Mary Ann Nicholl was killed, *Lloyd's Weekly London Newspaper* published a detailed account of the presumed murderer, a local creep who seemed to be quite well known:

> When the tragedy was first discovered on Friday the hapless females who haunt the East-end freely denounced a particular individual whom they style "Leather Apron." "Leather Apron" by himself is, it ap-

pears, quite an unpleasant character. He has ranged Whitechapel for a long time. He exercises over the unfortunates who ply their trade after 12 o'clock at night a sway that is based on universal terror. He has kicked, injured, bruised, and terrified a hundred of them who are ready to testify to the outrages... He carries a razorlike knife, and two weeks ago drew it on a woman called "Widow Annie" as she was crossing the square near London hospital, threatening at the same time, with an ugly grin and his malignant eyes, to do her harm... From all accounts he is five feet four or five inches in height, and wears a dark, close-fitting cap. He is thickset, and has an unusually thick neck. His hair is black, and closely clipped, his age being about 38 or 40. He has a small, black moustache. The distinguishing feature of his costume is a leather apron, which he always wears, and from which he gets his nickname.

Whoever Leather Apron was, he may or may not have been Jack the Ripper and may or may not have been responsible for some of the other six Whitechapel Murders. Today's criminologists and Ripperologists usually absolve Leather Apron of being the Ripper, due to the differences in style. Leather Apron swaggered around the streets making public threats, while the Ripper displayed a deep and very subtle pathology. It is the study of his evident psychosis that has driven Ripperology.

Part of this was a particular obsession with excising the kidneys and the uterus. It's often reported that Jack the Ripper must have been a doctor, due to the presumed need for medical knowledge in order to mutilate his victims. In fact this suggestion has very little support. Dr. Thomas Bond, one of several experts brought in by the police to examine the victims to gain insight into the killer, made the opposite suggestion, that the Ripper did not have medical knowledge:

... In each case the mutilation was implicated by a person who had no scientific nor anatomical knowledge.

In my opinion he does not even possess the technical knowledge of a butcher or horse slaughterer or any person accustomed to cutting up dead animals.

This was echoed by the murderer himself in one of his taunting letters mailed to investigators (September 25, 1888):

No luck yet, they say I am a Doctor now, ha ha.

Indeed his behavior was most un-doctorlike; as he also wrote in a letter the following month addressed "From hell" and accompanying a small parcel containing half of a human kidney preserved in a tiny bottle of spirits:

Mr Lusk. Sir I send you half the kidne I took from one woman prasarved it for you tother piece I fried and ate it was very nise I may send you the bloody knife that took it out if you only wate a whillonger. Signed Catch me when you can Mishter Lusk

Over the years, it's been suggested many times that a lot of Englishmen of the day had been involved in military campaigns in India and Africa. Various cultural traditions among certain Zulus and Indians would have become well known to British soldiers, including corpse mutilation. While this is intriguing, I found virtually no similarities between reported Zulu rituals and the things that the Ripper did. If the Ripper had been a soldier serving overseas at some point, I don't find that it had much to do with his murders.

In 1975, the *International Journal of Psychiatry in Medicine* published an interesting article that suggested the Ripper may have been medically traumatized as a child. The author surveyed the literature and found:

Katan writes that a child who has been frightened by a doctor may... attempt to discharge the anxiety he had experienced as a patient into his friend. We learn from the work of Buxbaum that a twelve-year-old boy who had been surgically traumatized at age six can... revel in fantasies of using a knife on someone's genitals. As Miller has pointed out, doctor-identification borne of a traumatic childhood surgical experience can remain well into adulthood and manifest itself in frightening fantasies of wanting to use a knife on the members of one's own family. Rose... shows that the senseless murder of an inoffensive stranger can represent the reenactment of consciously remembered, severely traumatic events in the murderer's childhood.

It's perhaps not entirely coincidental that the London stage adaptation of Robert Louis Stevenson's *The Strange Case of Dr. Jekyll and Mr. Hyde,* presented by the American actor Richard Mansfield, opened at the Lyceum theater on August 6, 1888, just three and a half weeks before the first of the Ripper's five murders. As the Ripper killed, the violent play ran to increasing public protest, depicting a man who transformed into a bloody killer. The play closed in late September, ostensibly due to negative publicity surrounding the Ripper's murders, and the proceeds from the final performance were donated to Whitechapel charities.

Whether the stage play may have inspired or even triggered the murderer is a possibility that, unfortunately, leads us nowhere. Maybe it did, maybe it didn't; the Ripper left us no evidence with which we might bolster speculation.

Leaving us with even less good evidence is the fact that the authenticity of *all* the Ripper's letters has been questioned. Handwriting analysts disagree on whether or not they match. Only one of the letters contains any information that would have been known to just a few people besides the real killer, and that's a letter written and received the same day that two of the five murders committed on the same night were discovered, and it boasts of them. But, the letter contains nothing that helps us learn the Ripper's identity. Beyond the few popularly reprinted letters were a grand total of over *six hundred* letters from "Jack the Ripper" received by various police and press agencies, and a number of people were actually arrested for writing hoax letters. Even the box with the kidney is shrouded in too much doubt to be certifiably authentic. As pointed out in 2008 in the *Nephrology Dialysis Transplantation* journal:

> *It appears beyond a reasonable doubt that the renal segment sent to George Lusk was human and this could be easily determined by morphological criteria in 1888. However, every medical student or person involved in the postmortem examination and/or with access to a mortuary could have obtained a human kidney. The kidney was preserved by alcohol suggesting that it may have been obtained earlier and may have been even part of a collection of anatomical specimens and was subsequently sent to Lusk after the press coverage of Eddowes' killing. There is even some, albeit indirect, evidence that the letters were written by journalists to keep the story boiling and to increase the circulation of the newspapers.*

The bottom line is that there is no widely accepted evidence suggesting an identity for Jack the Ripper. There are many hoaxed diaries, claims of royal conspiracies, Freemason plots, and all sorts of theories that are all either disproven or based on pure speculation. This lack of evidence is not surprising, for as the editors of the *Casebook: Jack the Ripper* website ably summarize:

> *Judging from modern cases, we now know that serial killers tend to be outwardly normal, even personable, and they attract very little attention to themselves. Jack the Ripper could very well have held regular employment, and even had a wife and children. In all likelihood the true murderer was never suspected by friends, family or coworkers, because he did not fit the profile of the Victorian raving lunatic.*

When you hear anyone claim to have finally solved the case of Jack the Ripper, you have very good reason to be skeptical.

REFERENCES & FURTHER READING

Barbee, S. "Introduction to the Case." *Casebook: Jack the Ripper.* Stephen P. Ryder & Johnno, 7 Dec. 2000. Web. 29 Mar. 2012. <http://www.casebook.org/intro.html>

Eckert, W. "The Whitechapel Murders: The Case of Jack the Ripper." *American Journal of Forensic Medicine and Pathology.* 1 Mar. 1981, Volume 2, Number 1: 53-60.

Editors. "Who Is Leather Apron?" *Lloyd's Weekly London Newspaper.* 31 Dec. 1888, Newspaper: 3.

Shuster, S. "Jack the Ripper and Doctor-Identification." *International Journal of Psychiatry in Medicine.* 1 Jan. 1975, Volume 6, Number 3: 385-402.

Sudgen, P. *The Complete History of Jack the Ripper.* New York: Carroll & Graf, 1994.

Whittington-Egan, R. *A Casebook on Jack the Ripper.* London: Wiley, 1975. 115.

Wolf, G. "A Kidney from Hell? A Nephrological View of the Whitechapel Murders in 1888." *Nephrology Dialysis Transplantation.* 11 Apr. 2008, Volume 23: 3343-3349.

46. I CAN'T BELIEVE THEY DID THAT: HUMAN GUINEA PIGS

A look at some of history's most famous scientists who experimented upon themselves.

Today we're going to take a look at one of the most interesting facets of science: Cases where researchers and volunteers have put themselves on the line for the sake of knowledge. There are a lot of cases where actual human experimentation is too risky. Even when volunteers are willing to undergo dangerous testing, ethics standards often prohibit their doing so. But nevertheless, throughout the history of science, a few maverick scientists and their volunteers have decided to walk the walk. They willingly, consensually, "took one for the team" and advanced our knowledge, with varying results. What follows are some of my favorite examples, the wildest and craziest cases that were undertaken by informed subjects with full knowledge of the possible consequences.

One of the best known such scientists was Col. John Stapp, a flight surgeon with the United States Air Force. Beginning in 1947, Stapp made dozens of runs in the famous rocket sled at what's now Edwards Air Force Base, testing the limits of hu-

Col. John Stapp aboard the *Gee Whiz*

man endurance of acceleration and deceleration. The sled named *Gee Whiz* was basically just a chair attached to rocket engines, wearing magnesium slippers that wrapped around the surface of railroad tracks to keep them from flying off. The *Gee Whiz* frequently destroyed equipment and the test dummy, Oscar Eightball. Stapp himself made 29 runs, and eventually allowed seven of his staff to make manned runs as well. Stapp's final run was at an even longer track in New Mexico on a much more powerful sled. He hit 632 mph (over 1,000 kph), shot past Joseph Kittinger flying a T-33 camera plane, and exploded into the lake of water that served as the brake. He sustained 46.2 Gs of force, equivalent to crashing your car into a brick wall at 120 mph. It burst all the capillaries in his eyes, turning them completely red, and gave him vision problems for the rest of his life. Stapp's earlier sled runs gave him two broken wrists, a broken collarbone, a sandblasted face, cracked ribs, several concussions, and he even had dental fillings shoot out of his mouth. The data he collected revolutionized aircraft seating, restraint, and ejection systems, and even provided the data that finally compelled Congress to require seatbelts in automobiles.

Then there's the case of Werner Forssmann, a young German doctor in residence, who, in 1929 at the age of only 25, thought that it might be possible to get a catheter into a patient's heart by inserting it into a vein in the elbow. He couldn't get permission from anybody to try it, and so he went maverick. He tricked the nurse who thought the procedure was going to be tried on her, and got

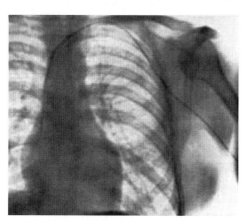

Werner Forssmann's X-ray proof

the catheter inserted into his own arm. He got it inserted most of the way, but then realized he couldn't continue without X-ray guidance. With the catheter still inserted, Forssmann walked downstairs to the hospital's X-ray department and stood in front of a mirror so he could see the catheter inside his own chest on a fluoroscope screen, a sort of live X-ray television. He completed the procedure successfully, inserting 60 cm (2 feet) of tube reaching all the way into his own right atrium, and took an X-ray photo as proof of his success. Although there was a lot of controversy about his action, and he ended up being unable to work in the field and had to become a urologist, the technique has become a foundation of cardiology. In 1956, Forssmann received the Nobel Prize in Medicine for his bold action.

One question that's never been completely solved is exactly what triggers the brain to go into panic mode and force you to gasp for air when you're trying to hold your breath. Suspecting that it might be the lack of contraction of breathing muscles, researcher Moran Campbell had his entire musculature paralyzed with intravenous curare, except for one forearm protected with a blood pressure cuff, with which he could signal. His heart, which is a different muscle type, would continue to beat but he could not breathe or move any part of his body except the one forearm. He was kept alive with a mechanical ventilator, which was then *switched off,* and observers watched to see what would happen. Breathless, Campbell signaled no discomfort for a full four minutes, when an attending anesthesiologist decided his blood CO_2 was too high and the ventilator was switched back on. A second volunteer underwent the same test with the same results. But then in 1989, a team at Harvard did a similar experiment, although they surreptitiously added CO_2 gas to the ventilated air, causing the subjects to signal distress immediately. The question remains unanswered, but the volunteers did not have to remain paralyzed, as the curare eventually wears off.

Another famous case was when the US Food & Drug Administration decided in 1903 to test various food preservatives, to see whether they could be used safely, and in what quantities. For five years, a group of young Department of Agriculture volunteers signed on for six-month terms during which they ate all their meals in a special dining room. At first, certain ingredients in the meals were prepared with one of the chemical preservatives being tested, but soon the program switched to serving regular, untreated meals and the men were given prescribed doses of various preservatives in capsule form. The program, established by Dr. Harvey Wiley, chief chemist of the FDA, was officially called the "hygiene table" but the press quickly gave it the name by which we know it today, the "poison squad". Both the volunteers and the press were given full knowledge of the program's details, and the volunteers were offered only the free meals as compensation.

The poison squad at a meal

Also braving poisons were J. S. Haldane, the Scottish physiologist, and his biologist son, J. B. S. Haldane, both mustachioed pipe-smokers in the grand tradition of Colonial Britain. Both were famous for their self-experimentation. Haldane the elder invented the gas mask that saved so many lives during World War I, largely as a result of spending time on the front lines analyzing poison gases, rushing to the scene whenever an attack was reported. He's also known for pioneering the use of the proverbial canary to detect dangerous gases in mines during the 1890s. Much twentieth century knowledge of the effects of various gases came from Haldane sealing himself inside a chamber and breathing them, noting any physiological and psychological effects.

Haldane the younger's destiny was sealed from the time he was a young boy; his father once educated him by taking him to a mine filled with firedamp gas and had him recite Shakespeare until he passed out. The younger became an expert on the effects of deep diving and the problems faced when rescuing submariners. He used a Royal Navy-funded compression chamber to subject himself and numerous volunteers to

J.B.S. Haldane

near-lethal diving conditions. He once went into a seizure during which he crushed a vertebrae. On other occasions, he suffered a collapsed lung, nerve damage to his spine resulting in several years of partial paralysis, countless nosebleeds, and frequently blew out his eardrums. His famous quote is "The drum generally heals up; and if a hole remains in it, although one is somewhat deaf, one can blow tobacco smoke out of the ear in question, which is a social accomplishment."

Jonas Salk is well known as the creator of the vaccine against polio. While soldiers fought overseas in World War II and Korea, this was the war being fought on the home front in the United States. When in 1955 it was publicly announced that Salk's new killed-virus vaccine was safe and effective,

Dr. Jonas Salk

and new incidences of polio dropped to practically zero, the celebration was massive. But to get there, at some early point in the research, the first volunteers would need to be injected with the virus itself. Killed virus, yes; but no one knew what the result would be. Then one day, at a press conference, Jonas Salk announced that the problem of finding volunteers had already been solved. He had been the first to receive it, followed by his wife and children. Salk and his staff watched and waited, and none of the first volunteers showed any sign of polio. It was a success.

One of Salk's fellow New York University alumni was US Army surgeon Major Walter Reed. In making possible the construction of the Panama Canal, Reed had to first eradicate yellow fever. He did this by tracing the disease vector to infected mosquitos, and proved it by having volunteers be deliberately bitten in rooms containing the annoying bugs. We thus learned for a fact how yellow fever was transmitted, and were finally able to take steps to reduce its proliferation by going after the mosquitos.

Dr. Max Joseph von Pettenkofer

Max Joseph von Pettenkofer was a doctor who studied hygiene, searching for ways to prevent the spread of cholera and typhus in countries that had unsanitary water supplies. One of his theories, which turned out to be wrong, was that the bacteria responsible for cholera was not dangerous unless it had incubated in warm, moist soil. To prove this, he ingested 1 cc of a cholera culture obtained from the stool of a recently deceased patient in 1892. Pettenkofer, and several of his students who followed the same experiment, contracted only mild cases and all recovered.

Just a few years earlier, in Peru in 1885, medical student Daniel Carrión was puzzling over Oroya fever. This infectious disease had been endemic in Peruvian populations since before recorded history. It wasn't known at the time, but it was caused by a bacteria transmitted by sand flies. Oroya fever has two stages: chronic and acute. In trying to find a link between the two, Carrión had himself injected with blood drawn from a boy with the chronic form of the disease. Although Oroya fever was only rarely fatal, Carrión developed the acute form, thus proving that both were caused by the same bacterium; and soon died. Twenty years later, those who followed his research identified the bacterium, and it's now easily treatable. Acute Oroya fever was renamed Carrión's Disease, and the young man who gave it all was declared a national hero. A stadium, a hospital, a university, and even a Peruvian province are named after him. And once a year, Peru celebrates *La Día de la Medicina Peruana*, the national Day of Peruvian Medicine, on October 5, the anniversary of Carrión's death.

Peruvian postage stamp featuring Daniel Carrión

Those who put themselves on the line don't always win their gambles. And so, the next time October 5 comes around, lift your glass and toast to those who took one for the team. Sometimes the circumstances made it impossible to do it any other way, sometimes there just wasn't enough information; but in all these cases, the scientists knew the risks and took the leap, and the state of our knowledge today is the better for it.

REFERENCES & FURTHER READING

Dendy, L., Boring, M. *Guinea Pig Scientists.* New York: Henry Holt & Co., 2005.

Denenberg, D., Roscoe, L. *50 American Heroes Every Kid Should Meet.* Brookfield: Millbrook Press, 2001. 98-99.

Garcia, G., Rodriguez, A. "Daniel Alcides Carrión y su aporte al conocimiento clínico de la fiebre de la Oroya y verruga peruana." *Cuaderno de Historia.* 1 Jan. 1995, Volume 80.

Jones, N. "Research in Exercise Physiology and Dyspnea at McMaster University." *Canadian Respiratory Journal.* 1 Oct. 2007, Volume 14, Number 7: 399.

Kevles, D. *In the Name of Eugenics: Genetics and the Uses of Human Heredity.* New York: Knopf, 1985. 122-128.

Lewis, C. "The "Poison Squad" and the Advent of Food and Drug Regulation." *U.S. Food and Drug Administration Consumer Magazine.* 1 Nov. 2002, Volume 36.

Spark, N. *A History of Murphy's Law.* Los Angeles: Nick T. Spark, 2003.

Trout, D. "Max Josef von Pettenkofer (1818-1901): A Biographical Sketch." *The Journal of Nutrition.* 1 Sep. 1977, Number 107: 1569-1574.

47. THE SIBERIAN HELL SOUNDS

Russian scientists are said to have drilled a borehole that broke into hell and released the screams of the damned.

There is one urban legend in particular that creeps out a lot of people. The story goes that sometime in 1989, Russian scientists in Siberia had drilled a borehole some 14.5 kilometers deep into the Earth's crust. The drill broke through into a cavity, and the scientists lowered some equipment to see what was down there. The temperature was about 1,100°C (about 2,000°F), but the real shocker was the sound that was recorded. They only got about 17 seconds of audio before the microphone melted, but it was 17 horrifying seconds of the screams of the damned.

Convinced that they'd heard the sounds of hell, many of the scientists quit the jobsite immediately, so the story goes. Those who stayed were in for an even bigger shock later that night. A plume of luminous gas burst out of the borehole, the shape of a gigantic winged demon unfolded, and the words "I have conquered" in Russian were seared into the flames. As a final touch of weirdness, medics were reported to have given everyone on site a dose of a sedative to erase their short-term memory. Beginning in 1989, the tale was broadly reprinted in smaller Christian publications, newsletters and the such, but was given hardly any notice by the mainstream media. Some evangelicals and Biblical literalists cited the incident as proof of the existence of a physical hell, an interpretation that seemed to be the consensus among the publications that ran the story. The story acquired the popularly conferred title of *The Well to Hell.*

The tale appeared just as the Internet began to rise, and as the legend grew, so did the number of debunks. By now the Internet is saturated with at least as many claims that either the audio or the story is false, as there are supporting it as fact.

The story's first appearance was in 1989 with its first large-scale publication by the Trinity Broadcasting Network. This Christian network has television shows in addition to a print newsletter, and they ran the story entitled "Scientists Discover Hell" in both their broadcast and print editions in late 1989. Shortly thereafter, they ran an expanded version of the story that included the newly reported detail of the devilish apparition coming up out of the borehole. Other Christian newsletters picked up the story, including *Praise the Lord* in February of 1990, and *Midnight Cry* in April of 1990.

But not everyone was on board. The first most obvious fact was that there was no such borehole in Siberia; however there was one on the Kola Peninsula in northwestern Russia, called the Kola Superdeep Borehole. Located only about 16 kilometers from Norway, the Kola borehole is about as far from Siberia as you can get and still be in Russia. Some researchers noted that the timing of the story was suspiciously close to that of an article in the August 1989 edition of *Science* magazine, titled "European Deep Drilling Leaves Americans Behind", which discussed the Kola project and a similar one in Germany. *Science* explained the purpose of the borehole:

> *To the Soviets, deep holes are not simply tools for testing geological theory. They expect more. One additional payoff is improved drilling technology. Another is insight into the deepest strata beneath known mineral resources... And there is always the allure of the Sputnik effect — the glory of having the deepest hole in the world.*

And so some competing Christian newsletters were more skeptical, noting not only the factual errors in the story, but also Trinity Broadcasting Network's lack of verifiable sources for their story. *Christianity Today* ran an article debunking the Well to Hell in July of 1990 (which we'll talk more about in a moment), as did *Biblical Archaeology Review* two months later.

Now, all of this happened without anyone ever hearing the alleged audio recording. Nobody ever presented one or broadcast one anywhere. It wasn't until twelve years later, in 2002, after the story had come and gone at least twice more in various tabloids, that a correspondent to Art Bell's radio program *Coast to Coast AM* emailed in an audio recording. The accompanying letter read as follows:

> *I just recently began listening to your radio show and could not believe it when you talked about the sounds from hell tonight. My uncle had told me this story a couple of years ago, and I didn't believe him. Like one of your listeners who discounted the story as nothing more than just a religious newspaper fabricated account. The story about the digging, the hearing of the sounds from hell, is very real. It did occur in Siberia. My uncle collected videos on the paranormal and supernatural. He passed away fairly recently... He let me listen to one of the audio tapes that he had on the sounds from hell in Siberia, and I copied it. He received his copy from a friend who worked at the BBC... Attached is that sound from my uncle's tapes.*

Bell then played the recording, and ever since then it's been widely circulated. To this day, the story is still reported from time to time, now with the supporting sound. It's become a firmly established urban legend.

But is it true? Not according to Rich Buhler, who was a radio host for *Christianity Today* in 1990, and who wrote the debunking article mentioned a moment ago. People had been calling into his show asking about the Trinity Broadcasting Network story, so Buhler and his staff did some digging. They worked backwards and followed all the threads they could to try and find whether there was a reliable original source. Here's what they found.

TBN had said on the air that their source was a Finnish newspaper called *Ammennusastia* which they described as a respected journal. An evangelical minister in Texas, R. W. Schambach, had come across it and sent it to TBN. It turns out that *Ammennusastia* was not a scientific journal at all, but was a small Evangelical Lutheran magazine that was published in Finland between 1974 and 1989. When Buhler contacted them to ask about the story, they reported that a staff member had written it from memory, having read the story in the daily Finnish newspaper called *Etelä-Suomen Sanomat* in a section that was for readers to contribute anything they liked, without verification. That reader had seen it in a Finnish paranormal newsletter called *Vaeltajat*. Buhler contacted *Vaeltajat* who reported that the story came from a reader who claimed to have seen it in a California newsletter published by Jewish Christians called *Jewels of Jericho* written by either an Alyde Carlsonin, Clyde Carlson, or Alyde Carlson. Nobody was ever able to track down *Jewels of Jericho* or verify its existence, so the trail went cold. I'm amazed that Buhler was able to follow the trail as far as he did. The whole chain was made of broken links: stories retold from memory, unverified sources, and no editorial scrutiny whatsoever. It is a nearly perfect example of a story without any solid foundation.

So if we can't verify any part of the story, where did that audio recording come from? It turns out that there is a popular explanation for it. Many Internet sites assert that it is a looped and layered version of an audio clip from the really terrible 1972 movie *Baron Blood*.

Personally I'm not convinced that the screams sound like the same ones; in fact, a side-by-side comparison serves mainly to convince me that *Baron Blood* is not the source of the audio. However, there's at least one really good YouTube video where a guy plays back selected samples from the Well to Hell audio proving that it is indeed looped.

Without any doubt, the Well to Hell audio played on the Art Bell show was created digitally by somebody looping and further processing some screaming sounds with a lot of background noise. That sound file, the only one known to exist from this story, is a hoax. There are zillions of recordings of screams and shouts and crowd noises for the hoaxer to have chosen from; whether or not he used *Baron Blood* is moot.

Further elements of the story have also been proven to be a hoax. In 1989, Norwegian teacher Åge Rendalen heard the original TBN broadcast while he was visiting California. Shocked at how gullible Americans were, he wanted to see how far it could be taken. He returned home, clipped a Norwegian newspaper article about a building inspector, and sent it to TBN along with a fake translation that added the new element of the figure of the devil coming up out of the borehole. Rendalen identified the photo of the building inspector as the Dr. Dmitri Azzacov (various spellings have been given) whom TBN had reported was the lead scientist of the project. TBN rebroadcast these startling new story elements without even bothering to do their own translation. Rich Buhler tracked down Rendalen who happily admitted his hoax, and all the details were laid out in the October 1990 issue of the *Secular Humanist Bulletin* newsletter.

And yet, the story continued to persist. The tabloid *Weekly World News* ran the story on April 7, 1992, but moved it to Alaska and added yet another new element of thirteen workers being killed when the devil came flaming up out of the hole. Sixteen years later on October 2, 2008, they ran it again on their online edition, changed it to an oil well, and added quotes from then-governor Sarah Palin and Vice Presidential candidate Joe Biden. In a subtle touch proving the tongue-in-cheek nature of their article, they located the site 400 miles north of Fairbanks, Alaska; which, to anyone with a map, places it squarely underwater in the Beaufort Sea.

So while we're able to prove that everything added to the original TBN story is a hoax, including the audio; all we can say about the original TBN story is that it was very poorly sourced and based on second and third hand accounts including personal recollections. We have no idea what the *Jewels of Jericho* used for their original source, or even if it existed at all. Certainly no such report of screams from hell ever made it into any legitimate geological publications. We know that all of its specifics are false: there is no such borehole in Siberia; drilling equipment can't operate at anywhere near the 1,100°C reported (the true maximum is less than 300°C), and neither can screaming human vocal cords.

Somewhere out there is a single anonymous person who first wrote into *Vaeltajat* with a story of fire and brimstone and eternal torment. That person could scarcely imagine how far the tale would go, and the extent to which researchers would puzzle over it more than twenty years later. The public is always hungry for a new urban legend, and always keener to accept that than to verify it.

REFERENCES & FURTHER READING

Brunvand, J. *The Baby Train.* New York: W. W. Norton, 1993. 105-108.

Buhler, Rich. "Scientists Discover Hell in Siberia." *Christianity Today.* 16 Jul. 1990, Newsletter: 28-29.

Cellania, Miss. "Human Oil (and Other Hoaxes)." *Neatorama.* Neatorama.com, 1 Nov. 2010. Web. 21 Apr. 2012. <http://www.neatorama.com/2010/11/01/human-oil-and-other-hoaxes/>

Editors. "Åge Rendalen." *Secular Humanist Bulletin.* 1 Oct. 1990, Newsletter.

Editors. "European Drilling Leaves Americans Behind." *Science.* 1 Aug. 1989, Volume 245, Number 4920: 816-817.

Mikkelson, B. "The Well to Hell." *Snopes.* Barbara and David P. Mikkelson, 17 Jul. 2007. Web. 21 Apr. 2012. <http://www.snopes.com/religion/wellhell.asp>

Schwarz, R. "The Dark Side of Eternity: The Siberia Recording." *Stranger Dimensions.* Stranger Dimensions, 18 Oct. 2011. Web. 21 Mar. 2012. <http://www.strangerdimensions.com/2011/10/18/the-dark-side-of-eternity-the-siberia-recording/>

48. PICNIC AT HANGING ROCK

Is this classic tale of the disappearance of a group of schoolgirls fact or fiction?

The year was 1900, the place southeastern Australia. A class of young women from a private boarding school, along with several chaperones, visited a scenic landmark in the country called Hanging Rock for a picnic. What happened that day has become a curious mixture of fact, fiction, and fantasy. The story holds that in a series of strange, almost dreamlike episodes, several of the girls went missing and were never seen again. Others went into inexplicable hysterics, and still others lost their memories of what had happened. By the end of the tale, two of the girls and one teacher were never seen again, and a third girl appeared to have taken her own life. The story became widely known in a 1967 novelization, which was soon made into a 1975 feature film, both titled *Picnic at Hanging Rock*. Today we're going to see if we can separate what really happened from what was dramatized, and study how the story went from one to the other.

Hanging Rock is a small volcanic formation about 50 kilometers northwest of Melbourne, Australia. In the midst of a broad green plain checkered with farmlands and vineyards stands a 100m tall bump of rock, the result of an ancient bulge of magma that rose and cracked and split apart. These spires and pinnacles and other formations are wreathed with forestry, and are popular with hikers, climbers, and photographers. The most famous formation is Hanging Rock, a large boulder that fell and jammed between two steep walls that you can now walk beneath and marvel up at. Mystery is a common theme at the park. The entry sign displays the tagline "Experience the mystery", and rangers report that tourists regularly mail back pieces of rock that they'd illegally removed, reporting that it had brought them bad luck. Some visitors even report haunted encounters with the ghosts of the missing girls.

Like many other landmarks in Australia, Hanging Rock had been a sacred ceremonial site for Aboriginals, and thus it carried with it a theme of mysticism. Australian author Joan

Hanging Rock

Lindsay was inspired by the place, and particularly by the juxtaposition of ancient spiritualism and modern colonial immigrants. Using this theme, she invented and wrote a novel, in only a single month, in which sophisticated upper class Europeans became trapped in a fanciful world in which they were, both literally and metaphorically, swallowed up by the ancient Earth. Yes, *Picnic at Hanging Rock* and the story that it tells are now, and have been ever since they were written, complete fiction. Our task today is to understand how and why a fictional story came to be perceived as fact.

The story goes that while exploring the rock, it's noted that three of the girls — Miranda, Marion, and Irma — along with their teacher Miss McGraw, have not been seen. A fourth girl, Edith, is in some sort of hysterics and reports that she saw the girls disappear into a cleft in the rocks, and also that she saw Miss McGraw inexplicably climbing the rock in her underwear. Men searched for several days, and finally found Irma alive four days later but with no memory of what had happened; strange, since Hanging Rock is small enough that it could easily be searched quite quickly. Where had Irma been? She's later demonized by other characters for failing to explain what had happened. At the college, a girl who was not allowed to attend the picnic takes her own life. The school's headmistress is later found dead, in unexplained circumstances, near Hanging Rock. No trace ever emerges of Miranda, Marion, or Miss McGraw.

Time was a motif in Lindsay's story. Flashbacks and missing time characterized her narrative, and the tragedy itself was foreshadowed by two characters whose watches stopped at exactly the same moment. Columnist Phillip Adams wrote that Joan Lindsay believed "times present, times past and times future coexist; that time isn't the simplistic continuum that most of us believe." She had no clocks in her home, and had titled her previous autobiographical book *Time Without Clocks*. Time was what separated the colonials from the Aboriginals. This culture clash is something that many Australians feel keenly, and it may well be responsible for why so many people have sought fact in the legend, to better confront their own place in an ancient land.

Lindsay did her own part to start the rumors that the book was based on fact. In the book's introduction, she wrote the following:

> *Whether Picnic at Hanging Rock is fact or fiction, my readers must decide for themselves. As the fateful picnic took place in the year 1900, and all the characters who appear in this book are long since dead, it hardly seems important.*

Near the end of the book she also referenced a newspaper article from 1914 about the disappearances. Said article never existed outside of the author's imagination, but few people fact check something like that.

And so from the moment the book came out, there was broad speculation that it may have been based on a true story. Independent researchers tried to verify some of its facts, such as whether the school named in the book (Appleyard College) was real or not. It wasn't, so researchers found no record of it; but all that does is leave the question open. Newspapers from the period were combed to see if any disappearances or deaths happened at Hanging Rock around 1900. One did; a young man slipped and fell to his death on New Year's Day in 1901. This lack of news coverage for what should have been a major headline should serve as circumstantial evidence that the event didn't actually happen, but enthusiasts keen on promoting the factual nature of a mysterious event are often hard to sway based on something so nebulous. In any case, the lack of a newspaper article doesn't prove one was never published; it just proves that such an article hasn't yet been found.

At some point, a rumor appeared that the local police station in Woodend had burned down sometime in the early 1900s, thus destroying any records of the girls' deaths that may have existed. There is (and was) an actual police station at Woodend, but there's no record that it ever burned down. The granddaughter of a police constable from Woodend, Richard Lawless, is reported to have phoned into a radio station and reported that her grandfather's theory is that the girls had fallen into a crevice which was then covered over by a boulder. If what she said is true — and there really isn't any way to know at this point — it proves nothing more than she believed her grandfather's tale.

What really stirred the pot was the 1980 publication of *The Murders at Hanging Rock* by Australian science fiction author Yvonne Rousseau. Although she prefaced her book with a statement that *Picnic at Hanging Rock* was fiction, she then went on to offer five possible explanations for the disappearances, since Lindsay had given none at all. Rousseau suggested:

1. That everything happened in some sort of parallel universe where time was slightly offset, thus accounting for why the bodies were never found, and explaining a major factual error in the original book. Lindsay had set her story on Saturday, February 14, 1900. However this date was actually a Wednesday.

2. A confusing suggestion that an alternate dimension was somehow involved.

3. They were all abducted by a UFO, which Rousseau suggested was consistent with Irma's amnesia.

4. A supernatural event of some kind must have taken place.

5. That it turns out to have been a conventional murder. The two teenage boys in the story, stable hand Albert and Michael, who was obsessed with Miranda, beat and raped and murdered Miranda, Marion, and Miss McGraw; but Irma escaped, having been beaten to the point of amnesia.

Although the first four explanations have never gained much traction outside of the New Age community (who still frequent Hanging Rock with crystals and robes), the murder story did take root. Among the many visitors who come to the rock today and ask the rangers about the mystery, it turns out that most of them have heard that the girls were murdered.

Finally, Joan Lindsay did eventually spoil her own party. As published, *Picnic at Hanging Rock* has seventeen chapters. But the first version submitted to her publisher had eighteen, and an editorial decision was made to cut out the final chapter which explained what happened with the girls, as it was felt that the mystery stood better without a solution. Lindsay left the manuscript with instructions that it be published three years after her death, and it was, in 1987. Titled *The Secret of Hanging Rock*, it turned out that all five of Rousseau's proposed explanations were wrong; and that what actually happened to the girls was so strange that the publishers were probably right to cut it out.

The whole chapter is ethereal and dreamlike. Miranda, Marion, and Irma are napping atop the rock. Fresh from her underwear rock climb, Miss McGraw (though she's never actually identified in the chapter) comes out of the bushes, still unclothed. In a gesture symbolic of burning the bridge from their previous culture, all four remove their corsets and throw them off the rock, but for some reason they hang suspended in mid-air. The girls follow a snake that descends into a strange hole. Miss McGraw magically transforms herself into a small half-crab, half-lizard thing and disappears down the hole after the snake. Marion does the same. Irma decides she doesn't want to and runs away. Finally, Miranda makes the same transformation, goes down the hole, and a rock falls and permanently seals it off.

It's unlikely that fans of the movie or the novel would have been satisfied with this ending, as up until that point, the story has all the makings of a proper mystery story. If the bizarre ending had been left intact, the book may have been a more fulfilling commentary from a metaphorical point of view, but it almost certainly would never have gained the popularity that the shortened version did. Mysteries crave solutions, and those mysteries that remain unsolved are the ones that perpetuate the craving. That's why Rousseau's book is probably always going to be more popular than the original author's own solution: it keeps the mystery going. And this is exactly why so many people want *Picnic at Hanging Rock* to be a true story: because it is a genuinely enduring and provocative mystery.

References & Further Reading

Lindsay, J. *Picnic at Hanging Rock.* Melbourne: F. W. Cheshire, 1967.

Lindsay, J. *The Secret of Hanging Rock.* Melbourne: Angus & Robertson Publishers, 1987.

McKenzie, B. *The Solution to Joan Lindsay's Picnic at Hanging Rock?* Self: Brett McKenzie, 1998. 1-9.

Rousseau, Y. *The Murders at Hanging Rock.* Fitzroy: Scribe Publications, 1980.

Stephens, A. "Hanging Out for a Mystery." *Sydney Morning Herald.* 13 Nov. 2008, Newspaper.

Watts, B. "The Mystique of Hanging Rock." *Joan Lindsay.* Pegasus Book Orphanage, 26 Oct. 2002. Web. 22 Apr. 2012. <http://www.bookorphanage.com/Joanlindsay.html>

49. THE SCIENCE AND POLITICS OF GLOBAL WARMING

Global warming is the poster boy for failed science communication. What went wrong?

Today we're going to open Pandora's box. We're going to point the skeptical eye at AGW, anthropogenic global warming, that part of climate change due to human actions. There's little dispute that climate change is real and natural; over the millennia, global climate patterns have gradually shifted. There's little dispute about natural global warming; we've been warming up for 150 years since the last lows of the Little Ice Age. But what there's plenty of dispute about — at least within the public if not within the scientific community — is AGW. How much are we humans causing or increasing the warming? What will the effects be of that increase if it exists? What, if anything, should we do about it?

Public understanding of AGW is all messed up, way more so than any other science, even more messed up than creation vs. evolution. The reason is obvious to everyone: It's never really been a science issue in the public's mind; it's a political issue. It's a political hot potato that has everyone on both sides of the aisle fired up and raging with conspiracy theories, fraud charges, end of the world scenarios, scandals and corruption. The result is that almost nobody in the public has a detailed understanding of the real science, yet almost everyone who follows the issue takes a side with great passion, either embracing AGW or dismissing it. What went wrong? How and why did this important science fly off the tracks and fall into the pit of politics?

This happened because AGW was never presented to the public the way science is; it was first presented as a political issue. The 1997 Kyoto Protocol, which made headlines when it went into effect in 2005, was the first time most people in the general public had any idea that global warming was a thing. Kyoto was a United Nations plan to reduce industrial emissions, but only in the wealthiest countries, and not at all in the biggest, poorest, dirtiest countries (China and India). It was deeply flawed scientifically, and effective really only as a slap in the face to the United States. Industrial powers, large on the political right, opposed it; environmental powers, largely on the left, supported it.

The second time the general public heard about global warming was also unscientific. Al Gore's 2006 movie *An Inconvenient Truth* was the first time that almost everyone had ever heard of global warming, and it was perceived as either gospel or as lies purely because of Al Gore's highly polarized position in the political world. Gore was a great champion of the Kyoto Protocol, and so was already perceived by conservatives more as an enemy of capitalism than as a defender of the environment. Whether anything he said was true or not, enough people went into the theaters predisposed to either embrace whatever he said or to reject whatever he said, that the actual content (even the whole subject itself) made little difference.

And that's exactly what got us where we are. By far the strongest predictor of a person's stance on global warming is his or her political affiliation. AGW is the poster boy of failed science communication. It is the perfect example of people embracing bad science because it agrees or disagrees with an ideology, either political or philosophical or ecological.

Think for a minute what would have happened if global warming had first been publicized by Stephen Hawking. I use Hawking as an example because he's really the *only* scientist who's known all around the world, and is universally respected and trusted. When Stephen Hawking tells you a science fact, you say "That's good enough for me." I would. And, like you, I haven't the slightest idea what Stephen Hawking's politics are. It's never occurred to me and I couldn't care less. Even though he's not at all a climate scientist, he's trusted by John Q. Public on matters of science, unlike Al Gore. If Stephen Hawking had been the one to make a movie about AGW instead of Al Gore, we might have no AGW denial at all in the world.

In this chapter, I am not going to try to convince anyone of anything by reading off lists of temperature measurements or CO_2 levels, or in fact give any facts and figures at all. Why not? Because that technique is a proven failure. Very few people have come to a conclusion on AGW that's fully independent of their ideology due to careful, impartial study of data. And for those who have, who knows whose interpretation of data they were studying? It's very easy to spin any data to show just about anything you want. A layperson has no way to know whether the facts and figures I'd give (or that anyone else might give) are in line with the AGW supporters, or with the AGW opponents. My goal here is to give a layperson the tools to come to an unbiased conclusion that truly is based on the best state of our current knowledge. For those of you on the right, this means setting aside distrust of the left; and for those of you on the left, it means setting aside distrust of the right.

Let me throw a couple of statements at you, one that you're likely to agree with, and one that you're likely to disagree with:

> **Statement A:** *Conservatives tend to reject AGW because it conflicts with their political agenda.*

> **Statement B:** *Liberals tend to embrace AGW because it fits nicely with their political agenda.*

Most of you (not all of you) strongly *disagree* with what I just said about your ideology, and you strongly *agree* with what I just said about the opposing ideology. Now about half of you are conservatives and about half of you are liberals, so *at least half of you have to be wrong.* If you have a truly open mind, you are open to the possibility that your own feelings on AGW are tempered with your ideology. It's probably fair to say that most people's are to some degree; it's perfectly fine to be satisfied when science facts fit nicely into your ideology. It's very difficult, and very rare, for *any* person's understanding of science to be completely divorced from their emotions or philosophies.

Name an expert you agree with on climate change. Almost nobody will name a real climate scientist; they'll name a communicator or pundit. In fact, name any climate scientist at all. None of them are famous; few of us have ever even heard one speak. Everything you know about climate science has been filtered through somebody else — probably someone you're predisposed to agree with.

So how can a layperson know what's probably right? As I always say on *Skeptoid,* go to our best scientific consensus and roll with that; you'll be right far more often than you'll be wrong. So this raises the obvious question: What is our best, real, apolitical, unbiased climate science consensus? Well, like it or not, the closest thing we have to that is the IPCC, the Intergovernmental Panel on Climate Change, established by the United Nations in 1988, a decade before the Kyoto Protocol was conceived and long before Al Gore ever heard of global warming. Now before you react, let's acknowledge some things. The IPCC is imperfect, and always has been. So is the prison system, so is the education system; every institution is a compromise of trying to please everyone. They're imperfect, but they're the best we're able to assemble. The criticisms are perfectly valid. *It's not necessary to regard the IPCC as a perfect institution to accept the scientific consensus that it provides.*

One facet that characterizes good science is that it evolves as our knowledge improves. In this way, the IPCC has produced four major versions of its assessment reports, and of as this writing is working on its fifth scheduled to be published in 2014. No responsible scientist has ever claimed that our current theories are absolutely correct and no further study is needed. The more work is done, the better our theories explain the observations. The tighter our predictions get. If you put any trust at all in the scientific method or in any branch of science, you know that our latest and best theories are just that. They're never final, they're never complete. It is not a weakness of the IPCC that they release revised assessment reports every few years; it's a strength. The constant publication of climate science articles in the best journals is not consistent with an old hoax, it's consistent with good science being done the way it's supposed to be done.

Is it possible that a synthesis of all the world's climate scientists is wrong about the science they spend their lives studying, and your favorite political commentators are right? Certainly it is. It's probably not very likely, but if it does turn out to be the case, then the synthesis will evolve in the direction that pans out through experimentation and observation, and future IPCC reports will be even closer to the facts.

Perhaps among the most flagrant mischaracterizations about AGW is that it's a long-debunked hoax or fraud, and that nobody takes it seriously anymore. This perspective reflects a total disconnect from current research. Fortunately, it's very easy to fix. Simply go to IPCC.ch (.ch is the top-level domain for Switzerland), click on the latest Assessment Report, and spend 10 minutes — or even 5 minutes — skimming the Summary for Policymakers. If you want to know the latest of the latest, go to the websites of the world's two leading scientific journals, *Science* and *Nature*. Search ScienceMag.org for the term "climate change" or look at *Nature's* sub-publication, *Nature Climate Change*. Take these minimal steps to inform yourself before telling anyone that AGW is a long-debunked hoax, or a fraud or a conspiracy. Such a perspective requires a deliberate disdain for current research.

One of George Carlin's most famous bits was titled *The Planet Is Fine*, in which he notes that no matter what kind of pollution humans produce, it really only affects those beings living there and not the planet itself. "The Earth isn't going anywhere," he says, "we are." Just where are we headed? Well, our best answers (so far) are there for you in black and white, if you're truly interested in knowing what they are.

REFERENCES & FURTHER READING

Angliss, B. "Serious Errors and Shortcomings Void Climate Letter by 49 Former NASA Employees." *Scholars & Rogues.* Scholars & Rogues, 25 Apr. 2012. Web. 6 May. 2012. <http://www.scholarsandrogues.com/2012/04/25/errors-shortcomings-void-nasa-climate-letter/>

Borenstein, S. "Skeptic Finds He Now Agrees Global Warming Is Real." *Yahoo! News.* Associated Press, 31 Oct. 2011. Web. 6 May. 2012. <http://news.yahoo.com/skeptic-finds-now-agrees-global-warming-real-142616605.html>

Brin, D. "The Navy, Russians, Shipping & Insurance Companies... and Climate Change." *Contrary Brin: Speculations on Science, Technology & the Future.* David Brin, 29 Mar. 2012. Web. 6 May. 2012. <http://davidbrin.blogspot.com/2012/03/navy-russians-shipping-insurance.html>

Douglas, P. "A Message from a Republican Meteorologist on Climate Change: Acknowledging Climate Science Doesn't Make You A Liberal." *Neorenaissance.* ShawnOtto.com, 28 Mar. 2012. Web. 6 May. 2012. <http://www.shawnotto.com/neorenaissance/blog20120329.html>

Laden, G. "HeartlandGate: Anti-Science Institute's Insider Reveals Secrets." *Greg Laden's Blog.* ScienceBlogs LLC, 14 Feb. 2012. Web. 6 May. 2012. <http://scienceblogs.com/gregladen/2012/02/heartlandgate_anti-science_ins.php>

Rosenau, J. "The Drama or the Soap Opera: the Future of Deniergate." *Thoughts from Kansas.* ScienceBlogs LLC, 21 Feb. 2012. Web. 6 May. 2012. <http://scienceblogs.com/tfk/2012/02/the_drama_or_the_soap_opera_th.php>

50. LEFT HANDED MYTHS AND FACTS

There are many popular anecdotes about how and why some people are left-handed, but the true facts are even more interesting.

Pretty much any Internet page discussing left-handedness facts and fiction contains lots of interesting little anecdotes, such as the history of how left-handed people have been treated or forced to convert, and how the word *sinister* comes from the latin for left. But a deeper study reveals that there's much more to this unusual trait, and plenty that we still haven't been able to fully explain.

About 8-10% of people are left-handed. That's a moderately interesting little factoid, but beneath it is a glorious patchwork of data and research, showing the variations of that number through history and across the continents and various demographics, and changes in our understanding of what it means. Why are 8-10% of people left-handed? How did they get that way? Does it really mean anything? Has this 8-10% always been a part of the human story?

One interesting way of assessing handedness from prehistoric times is to look at negative hand paintings in Paleolithic caves. The oldest of these come from western Europe, ranging from 10,000 to 35,000 years ago. The artist would hold a reed or straw in one hand, and blow paint at the other hand pressed up against the wall. Among 507 such hand paintings, 79 out of 343 (just about one quarter) are unambiguously of the left hand. This would seem to suggest that about three times as many people were left-handed then than today, but it turns out this is not the case. Research in France in 2002 gave this same task to college students, and found exactly the same distribution. Those same students, when asked to throw a ball, were 9% left-handed; and when asked to write, were 8% left-handed. Almost all who held the blowing tube in their right hand to paint a negative image of their left were normally right-handed, but among those who held it in their left hand, only a third were normally left-handed. This is exactly the same distribution observed among the Paleolithic artists.

Records go back even farther, as much as 400,000 years, when *Homo neanderthalensis* held meat in his teeth, pulled it tight with his left hand, and gripped a sharpened tool in his right hand to cut it off. Sometimes he slipped, and left marks on his front teeth that showed the direction of his cut.

The 8% number is not at all consistent worldwide. In what some researchers have termed "formal" countries, where handedness is strictly enforced in schools, we find very few left-handers: 3.5% in China, 2.5% in Mexico, 0.7% in Japan. In "informal" countries where little such redirection is given we find the highest numbers: up to a high of 12.8% in Canada. This redirected handedness is called learned handedness, but the more significant factors that we'll discuss include natural handedness (genetic or inherited), and pathological handedness, in which some type of brain injury or other condition produces the left-handedness.

Another reason that 8% is not always the result obtained in studies (a lot of times you'll hear the rounder number of 10%) is that there's a marked difference among both genders and age groups. Universally, males tend to be left-handed slightly more often than females; but what's really intriguing is that left-handers are found less often in older age groups. About 15% of 10-year-old children are left-handed, and this percentage steadily declines as people grow older. By the age of 90, there are virtually no left-handers left in the population. Women tend to live longer than men anyway; but when you add handedness, the difference becomes truly startling. According to a famous article published in 1991 in *Psychological Bulletin*, right-handed women have a life expectancy of around 77, but left-handed men only live to about 62. Conversely, left-handed women and right-handed men have nearly identical life expectancies, of right around 72. Overall, right-handers live 9 years longer than left-handers!

Why? Theories abound, but proof has been hard to come by. A popular suggestion has been that a lot of left-handers gradually switch over and become right-handed, due to the daily need to deal with the prevalence of right-handed tools, controls, and implements of all types; or possibly due to pressure from parents or teachers to switch for reasons such as ease of writing from left-to-right. However statistical analysis of the data shows that if change of handedness has any effect at all, it's very small. It turns out that the observation is indeed best explained by reduced longevity among left-handers.

The cause of this left-handed mortality is very difficult to find, since virtually no records exist. Death certificates don't include handedness, so there's really no data to analyze. We have theories, like accidents caused by left-handed living in a right-handed world. We do have observational data that supports the accident theory: in all five categories of life-threatening accident types (sports, work, home, tools, and driving), left-handers are from 1.2 to 1.8 times as likely as right-handers to suffer fatal accidents.

Certain health problems are also correlated with left-handedness, which probably also contributes to increased mortality. A 1988 survey found that in 30 of 33 publications, infants who had undergone birth stress were significantly more likely to be left-handed. Lower Apgar scores — a measure of a baby's overall condition at birth — have been clearly associated with left-handedness. A 1987 study found that more than a third of 4-year-olds who had been born prematurely were left-handed. Another found that more than half of children born with extremely low birth weights — a full 54% — were left-handed. In total, left-handers are twice as likely as right-handers to have had a stressful birth. Such births often result in long-term neurological damage. Hypoxia (the lack of oxygen to the brain) may well be one of the culprits. It's also been shown that mothers who smoke during pregnancy, which causes hypoxia to the fetus, are more likely to produce left-handed offspring.

Another very interesting observation is that certain immune deficiencies are more prevalent among left-handers. In some of these cases, elevated testosterone levels in the fetus is known to be the cause. The salient point about testosterone is that it's always found in higher levels among male fetuses, so elevated levels are more dangerous for males than for females. Recall that more males than females are left-handed. Elevated testosterone appears to slow the development of neurons in the left hemisphere of the brain, causing the right hemisphere to develop better, resulting in dominance of the left side of the body.

So the prevailing theory for why left-handers die out of the population is that a combination of factors put them at higher risk of impaired longevity: accidents, neuropathology, immune deficiencies, and other causes.

But this is not at all to say that all left-handers are born with some physical deficiency. No evidence has ever been found showing that *all* left-handers have anything in common. Heritability is a major factor: left-handed children are more likely to have left-handed parents. A lot of research has been published showing possible genetic markers, but so far there is certainly nothing as specific as a "left-handedness gene".

Moreover, it's certain that no genetic marker will ever be found that indicates left-handedness in every case. One of the reasons we know this for a certainty is something we've learned from twin studies, which are a great way to see genetic traits. If there were a genetic cause for handedness, we would expect identical twins, who are genetic copies of each other, to have concordant handedness, meaning that both are the same, either left-handed or right-handed. We would expect fraternal twins, who are genetically as different from one another as any other siblings, to have the same distribution of handedness as the general population. But, strangely, this is not what we see at all. It turns out that handedness among identical twins, fraternal twins, and the distribution among the general population are all equal. Identical twins are discordant, meaning one is left-handed and the other is right-handed, just as often as are fraternal twins, and as often as all siblings. This strongly suggests that genetics are not the overwhelming driver of handedness. There's no definitive answer as to how identical twins could end up discordant, but one leading theory is that is has to do with physical positioning within the womb. Twin fetuses usually face each other, and we've often observed their limbs adopt mirror-image positions. This behavior may require no more exotic an explanation than a simple economy of physical space available in the womb. Handedness is often predicted by which hand is most often held closest to the fetus' mouth, and this may indeed be the totality of the explanation for why identical twins are so frequently discordant.

The question everyone wants to know about handedness is what special aptitudes left-handers have. Are they more creative? Are they more likely to be successful? The most surprising thing to me that I found in my own literature survey is that there seems to be weak evidence for just about any aptitude you can think of, except the most commonly believed ones. It seems clear that there is no good evidence that left-handers are especially creative or that they're less talented at analytical functions, both of which would seem to indicate right brain dominance. But tasks are distributed over both halves of the brain, and even though a left-hander may rely on his right brain to control his left hand, aptitudes and mental tasks appear not to rely so much on that same type of hemisphere dominance.

Instead, much published research has examined just about every other aptitude. College educated left-handers seem to earn more money. They are more likely to be homosexual or to have gender identity problems. They tend to be better at geometry and spatial analysis. They are able to think outside the box better. The list goes on and on and on, but I found that nearly all such conclusions were based on a small number of small studies, many of which were contradicted by results published by others. So my conclusion, based on my own survey, is that there are not really any widely accepted aptitudes possessed by left-handers.

Sports and combat are a different matter. Most athletes and soldiers through history have been accustomed to going up against right-handed opponents, and a left-hander may bring his sword in or hit a ball from an unexpected angle. Today's top athletes do indeed have a larger distribution of left-handedness, especially in sports like baseball. But the statistics are clear that this is not due to any special talent above that of right-handers, but merely due to right-handers' greater physical difficulty in dealing with left-handed opponents.

So everybody hug a left-hander today. They don't have any special superpowers, but they are indeed at risk. It's the overriding issue of early mortality that is the focus of most of today's research, and rightfully so. And if you're in manufacturing, consider the importance of left-handed tools and controls, they're not just for convenience.

REFERENCES & FURTHER READING

Coren, S., Halpern, D. "Left Handedness: A Marker for Decreased Survival Fitness." *Psychological Bulletin.* 1 Jan. 1991, 109: 90-106.

Dellatolas, G., Tubert, P., Castresana, A., Mesbah, M., Giallonardo, T., Lazaratou, H., Lellouch, J. "Age and Cohort Effects in Adult Handedness." *Neuropsychologia.* 1 Jun. 1991, Volume 29, Issue 3: 255-261.

Hirsch, S. "Law and Mental Disorder." *Lancet.* 7 Apr. 1979, Volume 313, Issue 8119: 759–761.

Lord, T. "A Look at Hand Preference in Homo Sapiens." *The American Biology Teacher.* 1 Nov. 1986, Volume 48, Number 8: 460-464.

Medland, S., Perelle, I., De Monte, V., Ehrman, L. "Effects of Culture, Sex, and Age on the Distribution of Handedness: An Evaluation of the Sensitivity of Three Measures of Handedness." *Laterality*. 1 Jul. 2004, Volume 9, Number 3: 287-297.

O'Callaghan, M., Tudehope, D., Dugdale, A, Mohay, H., Burns, Y., Cook, F. "Handedness in Children with Birthweights Below 1000 g." *Lancet*. 16 May 1987, Volume 1, Number 8542: 1155.

Ross, G., Lipper, E., Auld, P. "Hand Preference of Four-Year-Old Children: Its Relationship to Premature Birth and Neurodevelopmental Outcome." *Developmental Medicine and Child Neurology*. 1 Oct. 1987, Volume 29, Number 5: 615-622.

Searleman, A., Porac, C., Coren, S. "Relationship Between Birth Order, Birth Stress, and Lateral Preferences." *Psychological Bulletin*. 1 May 1989, Volume 105, Number 3: 397.

Van Agtmael, T., Forrest, S., Williamson, R. "Genes for Left-Handedness: How to Search for the Needle in the Haystack?" *Laterality*. 1 Jan. 2001, Volume 6, Issue 2: 149-164.

Waldfogel, J. "Sinister and Rich: The Evidence that Lefties Earn More." *Slate*. The Slate Group, 16 Aug. 2006. Web. 12 May. 2012. <http://www.slate.com/articles/business/the_dismal_science/2006/08/sinister_and_rich.html>

Epilogue

I was amazed to learn that the water flowing from the springs at the bottom of Death Valley has been underground for thousands (if not tens of thousands) of years (source: USGS). It last landed as rain or snow in an area to the north some 40,000 square kilometers in size, and has been seeping through a vast aquifer over the course of all recorded human history. A raindrop that lands there today will not see the light of day for perhaps ten thousand years.

The typical visitor to Death Valley sees the water flowing from the spring, and probably looks up at the mountainside above and supposes "It must come from up there." They might spend a moment puzzling over why the water's flowing when it's midsummer, dry as a bone, and obviously no rain has landed up there anytime recently. And then, they probably never think of it again.

Would they be intrigued to hear of its marvelous underground journey, streaming through limestone tunnels, percolating through mineral deposits, dripping from stalactites, for uncounted millennia?

Some might. Many might not care.

I argue that taking the extra step to learn what's really going on, rather than simply accepting the apparently-true pop explanation, is always its own reward. Whether the subject is a basic Earth science like this example, or a classic ghost story in an old inn, or the latest money-making or health craze, what we can learn through study is always more exciting (and usually more true) that what we merely absorb through cultural osmosis.

Be the one with the hunger to know more. Be the one who second-guesses your group of friends. Be the one to question; but don't merely be contrarian, look instead to go deeper. I find that when I can raise an interesting question, it often piques the interest of others around me, simply because most people aren't used to wondering about things as much as they could. For many, it's an exciting new exercise.

All of the 50 short subjects in this book have caused me to raise my eyebrows and wonder. All of them prompted me to go deeper than I ordinarily might have, and all of them gave me a lasting reward that I often cherish.

I hope some of these chapters have done the same for you.

– Brian Dunning

INDEX

2012, 143, 157, 240

60 Minutes (news program), 80

9/11 attacks, 58, 185, 204, 209

911Truth.org, 204

absolutely converging series, 121, 122

acupuncture, 70, 73

acute stress response, 51, 52, 53, 54

Adams, John (mutineer), 215

Adams, Mike, 208

Africa, 233, 319

agency detection, 42

AgeOfAutism.com, 207, 208

Ahmadinejad, Mahmoud, 142

alcohol, 233, 234, 235, 236, 237, 321

aliens, 105, 106, 145, 146, 148, 149, 150, 289

Alt, Ernst, 24, 26

aluminum, 266

Am Fear Liath Mòr, 252, 254, 256

Ameranthropoides loysi, 302, 304, 305

American Civil Liberties Union (ACLU), 170

American Kennel Club, 227, 228

American Psychological Association (APA), 111, 113, 114, 115, 171

Amerika Bomber, 261

Ammennusastia, 334

Anderson, Anna, 97, 100

angular gyrus (brain), 87

Answers in Genesis, 205, 206

anti-Semitism, 139, 141, 142, 143

apocalypse, 240

approval voting, 192

Arnold, Larry, 66

Arrow, Kenneth, 189, 190, 193

astrobiology, 145

astrology, 31, 35, 108

Atlantis (mythical city), 264

Auschwitz, 260

Australian Vaccination Network, 206

autism, 207, 208

Bacon, Francis, 184

Bacon, Roger, 31, 32, 33

Bacon, Sir Francis, 184

bamboo, 43, 46, 128

barefoot, 71, 72, 73, 74

barefoot doctor, 71, 72, 74

Barefoot Doctor's Manual, A, 72, 73

barn owl, 198

Baron Blood, 334, 335

Beale Papers, The, 295, 298

Beale, Thomas, 295, 300

bears, 125, 150, 197, 219

Bell X-5, 261

Bell, Art, 333, 335

Ben MacDhui (mountain),
252, 253, 254, 255, 256,
257, 258
Bentley, John, 64, 65
Biden, Joe, 335
Big Pharma conspiracy, 106,
208, 287
bigfoot, 5, 6, 124, 126, 305
Black Hundreds, 139
Bligh, William, 211, 215
bokor, 90, 91, 92, 93, 94
Bolsheviks, 98, 99, 100, 141
Bond, Thomas, 318
Bowes-Lyon, Elizabeth
(Queen Mother), 159, 161,
162
Bowes-Lyon, Thomas, 160,
164
Boyle, Tom, 52, 53
brainwashing, 167, 168, 169,
170, 171, 172
breviograms, 183
Briggs, Benjamin, 233
British Broadcasting
Corporation (BBC), 65,
333
Broad Arrow Café, 38, 40
Brocken spectre, 254, 255
Brown Mountain Lights, 134
Bryant, Martin, 37, 38, 39,
40, 41
bryozoa, 291
Buhler, Rich, 333, 335
bulldog, 225
Bulwer-Lytton, Edward, 264
Bush, George H. W., 46, 192
Byrne, Peter, 126

caca de luna, 290
Cairngorms (mountain
range), 252, 253, 254, 258
Calanda, 17, 18, 19, 21, 22
Campbell, Moran, 325
cancer, 70, 245, 246, 249,
250, 282, 283, 284, 286,
287
carbon dioxide, 80, 81, 86,
219, 220, 325, 347
cardan grille, 32
Carlin, George, 349
Carrión, Daniel, 329
Carroll, Michael, 60
Carter, Jimmy, 169
Cartlidge, Doug, 81
Catholic church, 19, 23, 25,
111, 206
Central Intelligence Agency
(CIA), 46, 79, 80
Chapple, Steve, 79
Cheasapeake (tribe), 12, 14
chemtrails, 289
Cheney, Dick, 46
Chihuahua (dog), 227
China, 47, 69, 70, 71, 72,
127, 171, 201, 346, 352
Chinese Academy of Sciences,
70, 71
cholera, 328
chow chow, 229, 230
Christian, Edward, 214
Christian, Fletcher, 211, 212,
213, 214, 215, 216
Christian, Maimiti, 216
Christian, Thursday October,
215

Christianity, 112, 114, 139, 142, 174, 333, 334

Clinton, Bill, 169, 192

Coast to Coast AM (radio show), 333

coffee, 268

Coleman, Loren, 203

Coleridge, Samuel Taylor, 214

Collie, J. Norman, 253, 257

compact disc, 310, 311, 312, 313

computational stylistics, 186, 300

Conan Doyle, Arthur, 232

concussion, 324

Condorcet winner, 7, 189, 190, 192

Conservapedia, 203

conspiracy theories, 37, 39, 40, 41, 42, 46, 47, 56, 60, 61, 104, 105, 106, 108, 109, 138, 139, 142, 143, 152, 185, 204, 207, 208, 220, 234, 242, 264, 282, 283, 285, 286, 289, 321, 345, 349

conspiracy theorists, 37, 39, 40, 41, 42, 43, 47, 56, 60, 61, 104, 105, 107, 108, 109, 138, 143, 185, 207, 220, 242, 264, 282, 283, 285, 286, 289, 345

Cooper, Tim, 114

Croatan (tribe), 11, 14

Croatoan (island), 9, 11, 12, 14, 15

cryptography, 35, 299

Cryptomundo, 203, 204

cryptozoology, 126, 129, 194, 203, 304, 305

Cultural Revolution (China), 69

curare, 325

Currier, Prescott, 35

cyanobacteria, 290

datura, 93, 95

Daubert standard, 171

Davis, Wade, 91, 92

de Loys, François, 302, 303, 304, 305, 306, 307

de Loys' ape, 303, 304, 305, 306

de Vere, Edward (Earl of Oxford), 184, 185, 186

de Wet, Jacob, 164, 165

Dead Sea scrolls, 25

Death Valley, 241, 358

Death Valley, CA, 241

Defense POW/Missing Personnel Office (DPMO), 44, 47, 48

dehydration, 25, 221

Dei Gratia, 233

Delfin (USSR demolition), 80

delusional disorders, 107, 108

Demidova, Anna, 99

deprogramming, 167

detoxification, 202

DFS 228 (aircraft), 260

didi, 305

digital audio, 311, 312, 313

dimethyl sulfoxide (DMSO), 248, 249, 250

dinosaurs, 203

Diogenes the Cynic, 119

dissociative identity disorder, 25

DNA, 14, 92, 101, 102, 293

Doberman pinscher, 226

dog bites, 18, 225, 227, 228, 230, 231

dolphins, 76, 77, 78, 79, 80, 81

dopamine, 51

Dorey, Meryl, 206

Douyon, Lamarque, 92, 94

Doyle, Sir Arthur Conan, 232

Drake, Frank, 145

Drake, Sir Francis, 10, 145, 146

E-6B flight computer, 270, 271

Earhart, Amelia, 266, 267, 268, 269, 270, 271, 272

Earl of Strathmore, 159, 160, 164

earthquakes, 152

echolocation, 77

electrolysis, 262

Elizabeth I (Queen), 9, 159, 276

Elizabeth II, Queen, 101

epilepsy, 23, 25, 87

epinephrine, 51, 53

evolution, 32, 42, 104, 105, 128, 305, 345

Exodus, 112, 113, 114

Exodus International, 112, 113, 114

exorcism, 23, 24, 25, 26, 27, 28

Exorcist, The (movie), 27, 28

farming, 71

Fermi, Enrico, 145

fish, 79, 268, 293

Fitzgerald, James, 79, 80, 298

Folger, Mayhew, 214, 215

folie a deux, 277

Food and Drug Administration (FDA), 204, 205, 326

foot and mouth disease, 57, 58, 59, 61

Franklin, Benjamin, 195

Freemasonry, 139, 321

Freud, Sigmund, 111

frogs, 73, 92, 292, 293

fugu, 92

fungi, 290, 291

Gallager, Gerald, 269

gangrene, 17, 20

Gardner island, 267

Gasland (movie), 152, 154, 156

Gatty, Harold, 269, 270

Gee Whiz, 323, 324

Georgia Guidestones, 241

German shepherd, 226, 227, 228, 229, 230

ghosts, 6, 7, 41, 131, 134, 159, 160, 161, 162, 241, 242, 274, 275, 277, 282, 283, 284, 285, 286, 339, 358

Gigantopithecus, 124, 127, 128, 129

Gilgamesh (epic), 174

Glamis Castle, Scotland, 159, 162, 163, 164, 165

Glamis, monster of, 159, 160, 162, 165

global warming, 345, 346, 347, 348, 349

Glocke, Die, 262, 263

glycolysis, 51

gold, 14, 48, 242, 295, 296, 297, 298, 299

Golovinski, Matvei, 141

Google, 74, 156, 240, 241, 242, 243, 278, 289

Google Earth, 242, 278

Gorchynski, Julie, 245, 248, 249

Gore, Al, 192, 346, 347, 348

Great Leap Forward (China), 69

Great Michael (ship), 175, 176

Greenwood, Michael, 79, 80

Grey Man of Ben MacDhui, 252, 253, 254, 255, 256, 257, 258

Haiti, 90, 91, 92, 94, 95

Haldane, J. B. S., 326

Haldane, J. S., 326

Halliburton, 156, 157

Halloran, Walter, 27

hallucination, 280

Hanging Rock (park), 338, 339, 341, 342

Hatteras (tribe), 11, 14

Haukelid, Knute, 262

Hawking, Stephen, 346

Heinkel He-162 *Volksjäger*, 260

hepatitis, 245, 249

Hessdalen, Norway, 131, 133, 134, 135, 136

Heywood, Peter, 212, 214, 216

hibernation, 146

Hitler, Adolf, 24, 141

HMAV *Bounty*, 211, 212, 213, 214

Ho Chi Minh City, Vietnam, 45

Hodgins, Greg, 32

Hogg, James, 255

Hollywood, 43, 167, 186

Holmes, Sherlock, 232

Holocaust, 263

Holy Bible, 174

homeopathy, 202, 242

Homer, 118, 120, 121, 122

Homo neanderthalensis, 352

homosexuality, 110, 111, 112, 113, 114, 355

Hoodless, David, 269

Horten Ho-229, 260

Howland island, 267, 271, 272

Huffington Post, 202

hydraulic fracturing, 152, 153, 154, 155, 156, 157

hypothermia, 219, 220, 224

hypoxia, 83, 85, 87, 93, 354

Icke, David, 105, 106

Illuminati, 207

immune system, 205, 207, 209

Inconvenient Truth, An, 346

incorrupt corpses, 26

India, 81, 319, 346

InfoWars.com, 207

International Panel on Climate Change (IPCC), 348, 349

Internet, 41, 64, 81, 85, 195, 201, 202, 240, 263, 264, 292, 332, 334, 351

intervention, 20, 22, 115, 116, 169, 179

ionosphere, 132

iPhones, 241, 309

Iraq, 78, 175, 176

Iremonger, Lucille, 277, 278, 279

Israel, 138, 139, 142, 247

Jack the Ripper, 316, 317, 318, 319, 320, 321, 322

James, King, 12

Jamestown, Virginia, 12, 13

Japan, 57, 73, 74, 125, 267, 271, 272, 312, 352

Jersey devil, 194, 195, 196, 197, 198, 199

Jesuit (Catholic order), 32

Jews, 138, 139, 140, 141, 142, 143, 316, 334

Joly, Maurice, 140

Jones, Alex, 207, 208

Jourdain, Eleanor, 274, 275, 276, 277, 278, 279, 280

Kennedy, John Fitzgerald, 37

Kerry, John, 46

ketamine, 87

Key West, 79

King, Kenneth, 61

Kittinger, Joseph, 324

Korea Consumer Protection Board, 219, 224

Korean fan death, 218, 219, 220, 221, 222, 223, 224

Krafft-Ebing, Richard von, 111

Krushevan, Pavel, 140, 141

Kundalini yoga, 66

Kyoto Protocol, 346, 348

Labrador retriever, 227, 228, 229

Landkreuzer P.1000 *Ratte*, 261

Landkreuzer P.1500 *Monster*, 261

Laos, 48

Lawrence Livermore National Laboratory, 248

Leather Apron, 317, 318

Leeds, Daniel, 195, 196, 197

left handedness, 351, 352, 353, 354, 355, 356

Lépekhine, Mikhail, 141

Ley, Willy, 264

Lindbergh, Charles, 270

Lindsay, Joan, 339, 340, 342

Lippisch P.13a, 260

Lippisch, Alexander, 260

Lockheed Model 10 *Electra*, 266, 267, 271

Lockheed T-33 *Shooting Star*, 324

Lockheed U-2, 260

lottery voting, 190, 191

Lucifer Project, 24

MacFadden, Fred, 196, 197, 198

Malvo, Lee Boyd, 171

Manhattan Project, 261

Mao Zedong, 69, 70, 71, 72, 74

Marfa Lights, 134, 136

Marlowe, Christopher, 183, 184

Mary Celeste, 232, 233, 234, 235, 236, 237

mass hysteria, 247

McCain, John, 46

Melbourne, Australia, 338

Mercola, Joe, 204, 205

Mercola.com, 204, 205

mercury (element), 207, 208

Mercury program, 50

Messerschmitt Me-163 *Komet*, 260

Messerschmitt Me-262 *Schwalbe*, 260

Messerschmitt P.1101, 260

Messori, Vittorio, 19

metabolism, 51

methane, 66, 152, 153, 154, 155

Michel, Annaliese, 23, 24, 25, 26, 27, 28

Min Min light, 134

mineralogy, 299

mirage, superior, 134, 241

Moberly, Anne, 274, 275, 276, 277, 278, 279, 280

Montandon, Georges, 304, 305, 307

Montauk monster, 56, 61

Moody, Raymond, 84, 88

Morehouse, David, 233, 234, 237

Morriss, Robert, 296, 297, 298, 300

Mozart, Wolfgang Amadeus, 185, 186

Muhammad, John Allen, 172

multiple personality disorder, 25

Mustang (aircraft), 314

myxobacteria, 290

Nader, Ralph, 192

Narcisse, Clairvius, 91, 92, 93, 94, 95

National Association for Research & Therapy of Homosexuality (NARTH), 113, 114

NaturalNews.com, 208, 209

Nazi Germany, 139, 259, 260, 261, 262, 263, 264, 265

near death experiences, 83, 84, 85, 86, 87, 88

necrosis, 245

Nepal, 126, 129

new age, 119, 169, 264, 342

New World Order, 207, 209

Newcombe, Phyllis, 67

Nickell, Joe, 197, 298, 300

Nikolaevna, Alexei, 99, 101, 102

Nikolaevna, Anastasia, 97, 99, 100, 101, 102

Nikolas II, 97, 99, 101

Nikumaroro island, 267, 268, 269, 271

Nixon, Richard, 58

Nobel Prize, 189, 325

Noonan, Fred, 267, 270

North Carolina, 10, 134

Nostoc (cyanobacteria), 290

nursery effect (elections), 191

Nyquist rate, 310

Obama, Barack, 107

Occam's razor, 22, 83
occult, 257, 259, 263
Ogilvie clan, 163
Okhrana, 141
Olmsted, Dan, 207, 208
Operation Gunnerside, 261
Operation Homecoming, 44
organic food, 33
Oroya fever, 329
Oscar Wilde, 111
Osman-Hill, William, 127
oxygen, 83, 87, 95, 208, 219, 220, 222, 248, 249, 354
Pacific Ocean, 266, 267
Palin, Sarah, 335
Pangboche (monastery), 126, 129
panspermia, 289
parchment, 30, 32, 33, 309
parietal lobe (brain), 87
Parmenides, 118
pathogens, 58, 292
Patrick, Ted, 170
pearl diving, 268, 269
Pellicer, Miguel Juan, 17, 18, 19, 20, 21, 22
Penthouse (magazine), 79, 80
People (magazine), 250
Perkins, Marlin, 129
Perot, Ross, 192
Petit Trianon, 274, 278
Pettenkofer, Joseph von, 328
phencyclidine (PCP), 54
Picnic at Hanging Rock, 338, 339, 340, 341, 342, 343
piezoelectric effect, 132, 133

Pioneer (spacecraft), 147, 148
pit bull, 225, 226, 227, 229, 230, 231
Pitcairn islands, 211, 212, 214, 215, 216
Planck length, 120
plasma, 132, 133, 135, 249
Plato, 118
Plum Island Animal Disease Center, 56, 57, 58, 59, 60, 61
Plutarch, 178
plutonium, 261, 262
Pocahontas, 12
poison squad, 326
polio, 327
polytheism, 91, 108
poodle, 227
Port Arthur, Tasmania, 37, 38, 40, 41, 42
Powhatan (tribe), 12, 13, 14
primatology, 127, 128
prisoners of war, 43, 44, 46, 48
PrisonPlanet.com, 207
Protocols of Zion, 139, 140, 141, 142
psychopathology, 107, 108
pufferfish, 92, 93
Putney Swope (film), 188
pyrotron, 66
Quinn, David Beers, 14
Rachkovsky, Pyotr, 141
radiofrequency identification (RFID), 209
radiometric dating, 203
Raleigh, Sir Walter, 9, 10, 12

Ramirez, Gloria, 245, 246, 247, 248, 249, 250

random ballot, 190

range voting, 191, 192

rauding, 345

reasoning errors, 282, 287

Red Flag, 70, 71

Reed, Walter, 328

Reeser, Mary, 64, 65

Rendalen, Åge, 335

Rennie, John A., 256, 257

Renz, Arnold, 24, 26

reptoids, 106

rinderpest, 58

ripperology, 317, 318

RMS *Titanic*, 176

Roanoke, 9, 10, 11, 12, 13, 14, 15

Robert Louis Stevenson, 320

robots, 148, 149, 150

Rolfe, John, 12

Romanov family, 97, 99, 100, 101, 102

Rottweiler, 230

Rousseau, Yvonne, 341, 342, 343

Russell, Bertrand, 120

Saari, Don, 190

Saffin, Jeannie, 67

Salieri, Antonio, 185

Salk, Jonas, 327

Sartori, Penny, 86, 88

schistosomiasis, 71, 72

schizophrenia, 25

Schlafly, Andrew, 203

Scientology, 168, 171

scopolamine, 93

sea lions, 77, 78

seaquakes, 235

Search for Extra-Terrestrial Intelligence (SETI), 145

Sella, Andrea, 237

Serpent and the Rainbow, The (book), 91

sexual orientation change efforts (SOCE), 110, 112, 113, 114, 115

Shakespeare, William, 181, 182, 183, 184, 185, 186, 327

Shark Dart, 80

sheeple, 209

Shipton, Eric, 125

Siberian hell sounds, 331, 333, 334, 335

Silberschlag, Johann, 254

Silbervogel (space plane), 261

silver, 262, 295, 297, 299

skeletons, 10, 94, 234, 269, 270

skepticism, 17, 19, 69, 239, 282

Slick, Tom, 126, 129

Smith, John, 12

Society for Psychical Research, 278, 279

sociogenic illness, 246, 247, 248, 250

sonar, 77, 79

Soul City, 45, 46

Soviet Union, 46, 79, 80, 98, 140, 332

Spetsnaz, 80

Spitzer, Robert, 113

spontaneous human combustion, 63, 64, 66, 67, 68

Sputnik (satellite program), 260, 332

SS *Norwich City*, 269

Stanley, William (Earl of Derby), 184

Stapp, John, 323

star jelly, 289, 290, 291, 292, 293

Strachey, William, 12, 13

Strand, Erling, 131

Strange Case of Dr. Jekyll and Mr. Hyde, The, 320

stress, 25, 51, 53, 112, 177, 247, 353

sugar, 92, 95, 205

Superman (various iterations), 178

supersonic flight, 260

Symbionese Liberation Army, 169

Tahiti, 211, 212, 214, 215

Tejera, Enrique, 306, 307

Teodorani, Massimo, 132, 133, 134, 136

terrier, 225, 229

Tessarakonteres (ship), 178

tetrodotoxin, 92, 93, 94, 95

Tewnion, Alexander, 253, 256

thermal lensing, 134

Three Mile Island, 66

TIGHAR, 267, 268, 269, 270, 271

tobacco, 12, 327

toxic lady, 246, 247, 250

traditional Chinese medicine (TCM), 69, 73, 128

Trinity Broadcasting Network (TBN), 332, 333

Trondheim Airport (TRD), 135

Tschaikovsky, Anna, 99

Turing, Alan, 111

typhus, 328

UFOlogy, 131

unidentified flying objects, 131, 135, 264, 275, 342

United States Air Force, 323

United States Department of Agriculture (USDA), 57, 58, 59, 61, 326

United States Department of Homeland Security (DHS), 58, 59

United States Environmental Protection Agency (EPA), 156, 157

United States Geological Survey (USGS), 358

United States Navy, 35, 77, 78, 79, 80, 81, 327

United States Navy Marine Mammal System, 77, 78, 81

United States Navy Shallow Water Detection System, 78

Uranverein, 261, 262

USCGC *Itasca*, 267, 271, 272

USS *Lexington*, 272

V-2 (*Aggregat-4* rocket), 260, 261

V-2 (rocket program), 260, 261

vaccines, 7, 58, 209, 327

Valencia, Spain, 17, 19, 20, 21

Vatican, 26

Ventura, Jesse, 56, 60, 61

Versailles, 274, 275, 276, 277, 278, 279

Vietnam, 43, 44, 45, 46, 48, 78, 80

Vikings, 54

vinyl records, 309, 312, 313, 314

viruses, 289, 292, 328

vitamins, 282, 283, 284, 286, 287

vodou, 90, 91, 95

voting, 7, 188, 189, 190, 191, 192, 193

Voyager (spacecraft), 147, 148

Voynich manuscript, 30, 32, 33, 35, 36

Vril Society, 264, 265

Ward, James B., 295

waterspout, 235

Weekly World News, 335

Wentworth Day, James, 161, 162, 163

West Nile virus, 60

Westminster Abbey, 181

whales, 77, 78, 79

Whinnery, James, 87

White, John, 10, 11, 13, 15

Whitechapel murders, 316, 317, 318, 320

wick effect, 65, 66, 68

Wikipedia, 177, 203, 243

Wiley, Harvey, 326

Wilson, Sam (Uncle Sam), 81

witchcraft, 162

Witkowski, Igor, 263

World War I, 45, 57, 97, 271, 302, 326, 327

World War II, 45, 57, 271, 327

wormholes, 147

Wunderwaffen, 259, 262, 265

Xerxes (king), 139

x-rays, 324, 325

Yale University, 32, 59

Yeltsin, Boris, 46, 47

yeti, 7, 124, 125, 126, 127, 128, 129, 257

young earth creationism, 203

YouTube, 283, 334

Yurovsky, Yakov, 98, 99, 101

Zaragoza, Spain, 17, 18, 19, 21

Zatsiorsky, Vladimir, 52, 53, 54

Zeno of Elea, 117, 118, 119, 121, 122

Zheng He, 178

Zielgerät ZG-229 *Vampir*, 260

Zionism, 138, 139, 142, 143

Znamya (newspaper), 139

zombie powder, 92, 93, 94, 95

zombies, 27, 90, 91, 92, 93, 94, 95

zooids, 291

Zuniga map, 13